T0135489

Lienhard Pfeifer

Pedestrian Detection Algorithms
using Shearlets

Logos Verlag Berlin

Bibliografische Information der Deutschen Nationalbibliothek

Die Deutsche Nationalbibliothek verzeichnet diese Publikation in der
Deutschen Nationalbibliografie; detaillierte bibliografische Daten sind
im Internet über http://dnb.d-nb.de abrufbar.

ISBN 978-3-8325-4840-7

Logos Verlag Berlin GmbH
Comeniushof, Gubener Str. 47,
D-10243 Berlin
Tel.:   +49 (0)30 / 42 85 10 90
Fax: +49 (0)30 / 42 85 10 92
http://www.logos-verlag.de

Dissertation

Pedestrian Detection Algorithms using Shearlets

Lienhard Pfeifer

2018

Pedestrian Detection Algorithms using Shearlets

Dissertation

zur

Erlangung des akademischen Grades

Doktor der Naturwissenschaften

(Dr. rer. nat.)

vorgelegt

dem Fachbereich Mathematik und Informatik

der

Philipps–Universität Marburg

(Hochschulkennziffer: 1180)

von

Lienhard Pfeifer

geboren am 20. August 1985

in Gladenbach

Erstgutachter: **Prof. Dr. Stephan Dahlke**, Philipps-Universität Marburg

Zweitgutachter: **Prof. Dr. Bernd Freisleben**, Philipps-Universität Marburg

Tag der Einreichung: 11.9.2018

Tag der mündlichen Prüfung: 9.11.2018

For Zoe, Sam and Eva.

Acknowledgment

First of all, I would like to thank my supervisor Prof. Dr. Stephan Dahlke for his constant support and a lot of valuable advice. Moreover, I thank him for accepting me as an external PhD student. But mostly, I thank him for his guidance and encouragements never showing any doubt that my work will be successful. I also thank Prof. Dr. Bernd Freisleben for agreeing to be a referee of this thesis.

I would like to thank Matthias Gemmar for his enormous and never ending support for me and my project at ITK Engineering GmbH. I thank him for giving me always constructive feedback, helpful advice and good ideas. Especially, I thank him for promoting my project at ITK and for always believing in the benefit of it.

I want to thank the management of ITK Engineering, namely Michael Englert and posthumously Dr. Helmut Stahl for their decision to fund my thesis. In addition, I thank my supervisors at ITK, especially Christian Hötterges and Jens Weber, for backing this decision and for supporting me and my project.

I also thank my colleagues at ITK for fruitful discussions as well as providing distraction by enjoyable talks apart from work. A special thank goes to Max Rasumak and our former student Sven Heuer for their input into this project. I thank Dr. Sören Häuser for providing me with a lot of insights concerning the design of shearlets and digitalization, respectively the implementation of them. Furthermore, I thank Dr. Philipp Petersen from the TU Berlin for his help on theoretical aspects of compactly supported shearlets. For proofreading I thank Elisabeth Fiedler, Alexandra and Joseph Marmion, Dr. Sören Häuser and Max Rasumak.

I want to thank my family and friends for their interest in my work, their support and for delivering pleasant times. In particular, I thank my father Roland for awakening my interest in natural sciences and my deceased mother Sabine for her loving education. A special thank goes to my son Sam for clearing up my mind by playing around and to my daughter Zoe for being such a sunshine. Finally, I thank my wife Eva for her patience, her love, her support, for bringing our two beautiful children into the world, for being a wonderful mother for them and above all for sharing the best moments with me.

Abstract

In this thesis, we investigate the applicability of the shearlet transform for the task of pedestrian detection. Due to the usage of in several emerging technologies, such as automated or autonomous vehicles, pedestrian detection has evolved into a key topic of research in the last decade. In this time period, a wealth of different algorithms has been developed. According to the current results on the Caltech Pedestrian Detection Benchmark [31], the algorithms can be divided into two categories. First, application of hand-crafted image features and of a classifier trained on these features [28, 99, 130]. Second, methods using Convolutional Neural Networks in which features are learned during the training phase [9, 32, 127]. Our aim is to study how both of these types of procedures can be further improved by the incorporation of a framework for image analysis which has a comprehensive theoretical basis.

We choose the multi-scale framework of shearlets since it guarantees a unified treatment of the continuum and the digital world [74]. In theory, shearlets provide optimally sparse approximations of certain image models [52, 76] and the ability to characterize edges in images [53, 78]. Moreover, they have been successfully applied practically in several image processing tasks, for example denoising [35, 86] and edge detection [125].

We adapt the shearlet framework according to the requirements of the practical application of a pedestrian detection algorithm. The particular shearlet design consists of two parts. First, the usage of a specific mother shearlet that can precisely locate structures in images. Second, the setup of a shearlet system allowing a uniformly distributed directional analysis and flexible adjustment of the shearlets used per scale. The last point is useful to gain control over the space size of the image features used for pedestrian detection. We show that our shearlet system forms a frame for $L^2(\mathbb{R}^2)$ and provide the conditions required for it.

Next, we examine the capability of our shearlet design for edge detection from a theoretical point of view. We show that the designed shearlets can characterize edge points in $\mathbb{R}^2$ and their type by the decay rates and the limits of the shearlet transform for decreasing scales. In contrast to the existing literature, we derive decay rates depending on the degree of anisotropy α. For the special case of $\alpha = 1/2$, we find explicit limits of the shearlet transform which has not been achieved with recently used shearlets. Furthermore, we show that a shearlet mother function with just one vanishing moment is required for our theoretical results. Finally, we illustrate that this requirement is in harmony with the observations made in the practical application of an edge detection algorithm with shearlets.

Given our specialized shearlet design, we define meaningful, hand-crafted image features based on the shearlet transform. Meaningful in a sense that the features provide rich information about an object's structure. The consideration of shearlets has several advantages compared to other image features. At first, they provide a sparse representation of the image, i.e. the result of the shearlet transform only has magnitude with considerably high absolute value at pixels that correspond to edge points. Second, due to the multi-scale framework of shearlets, the structure of objects can directly be investigated at different scales. We show that our shearlet features provide the best results of hand-crafted image features in the Caltech benchmark.

Currently, all best performing algorithms in the Caltech benchmark are using CNN models. Therefore, we analyze the capability of shearlets to further improve CNNs. We integrate shearlet filters in the first layer of a CNN, since several learned filters of this layer show a similarity to orientation-selective edge detection filters [72]. We show that we can improve classification and detection results with the integration of shearlets compared to the original networks when trained on the same data. Furthermore, we find that a training on comprehensive data sets such as ImageNet [24] is required in order to achieve the detection results of the leading algorithms. We consider this finding as indication for the immense power of data for deep learning algorithms.

One main application area of pedestrian detection is located in the automotive domain. The usage in current collision warning and intervention systems [40] as well as in future systems of autonomous driving requires algorithms to be runable on embedded devices. Therefore, we port our base pedestrian detection algorithm to a marketably priced embedded target. By a careful software design and runtime optimization we are able to run our algorithm with a useful frame rate of 10 fps.

Zusammenfassung

Die vorliegende Arbeit untersucht die Eignung von Shearlets für die Aufgabe der Fußgängererkennung. Durch die mögliche Anwendung der Fußgängererkennung in diversen neuen Technologien, wie beispielsweise dem automatisierten oder autonomen Fahren, hat sich das Thema im letzten Jahrzehnt zu einem Schlüsselthema der Forschung entwickelt. In dieser Zeitperiode wurde eine Vielzahl verschiedener Algorithmen entwickelt. Nach den aktuellen Resultaten des Caltech Pedestrian Detection Benchmarks [31] können die Algorithmen in zwei Kategorien unterteilt werden. Zum einen, die Anwendung von handgefertigten Bildmerkmalen und eines Klassifikators, welcher auf diese Bildmerkmale trainiert ist [28, 99, 130]. Zum anderen, Methoden mit Anwendung von Convolutional Neural Networks (CNNs), in denen Bildmerkmale während der Trainingsphase gelernt werden [9, 32, 127]. Unser Ziel ist es zu untersuchen, in welcher Weise beide Typen von Algorithmen durch die Einarbeitung eines Frameworks mit umfassender theoretischer Grundlage verbessert werden können.

Wir wählen hierzu das Multiskalen-Framework der Shearlets, da es eine einheitliche Behandlung der kontinuierlichen als auch der digitalen Welt garantiert [74]. In der Theorie haben Shearlets optimal sparse Approximationen bestimmter Bildmodelle [52, 76] ermöglicht und haben die Fähigkeit Kanten in Bildern zu charakterisieren [53, 78]. Des Weiteren wurden Shearlets erfolgreich für diverse Bildverarbeitungsaufgaben, wie z.B. Denoising [35, 86] oder Kantendetektion [125], eingesetzt.

Wir passen das Shearlet Framework entsprechend der Anforderungen der praktischen Anwendung einer Fußgängererkennung an. Das entsprechende Shearlet Design besteht aus zwei Teilen. Zum einen, die Verwendung spezieller Shearlet Mutterfunktionen, welche Bildstrukturen präzise lokalisieren können. Zum anderen, das Setup eines Shearlet Systems welches eine gleichmäßig verteilte, gerichtete Analyse und eine flexible Anpassung der Shearlets pro Skala zulässt. Der letzte Punkt ist nützlich um eine bessere Kontrolle über die Raumgröße der Bildmerkmale zu erhalten. Wir zeigen, dass unser Shearlet System einen Frame für $L^2(\mathbb{R}^2)$ bildet und liefern die dafür benötigten Bedingungen.

Im Folgenden untersuchen wir die Fähigkeit unseres Shearlet Designs zur Kantendetektion aus theoretischer Sicht. Wir zeigen, dass mit unseren Shearlets Kantenpunkte in $\mathbb{R}^2$ und deren Typ durch Abfallraten und Grenzwerte der Shearlet Transformation für abnehmende Skalen charakterisiert werden können. Im Gegensatz zur aktuellen Literatur erhalten wir Abfallraten abhängig von dem Grad der Anisotropie α. Für den Spezialfall $\alpha = 1/2$ ermitteln wir explizite Grenzwerte der Shearlet Transformation welches mit bisher verwendeten Shearlets nicht erreicht wurde. Zudem zeigen wir auf, dass wir eine Shearlet Mutterfunktion mit lediglich einem verschwindenden Moment benötigen um unsere theoretischen Resultate zu erhalten. Schließlich stellen wir dar, dass diese Anforderung in Einklang mit den Beobachtungen bei der praktischen Anwendung eines Kantendetektionsalgorithmus mit Shearlets steht.

Basierend auf unserem speziellen Shearlet Design definieren wir aussagekräftige, handgefertigte Bildmerkmale. Aussagekräftig in dem Sinne, dass die Bildmerkmale reichhaltige Informationen über Objektstrukturen liefern. Die Betrachtung von Shearlets hat mehrere Vorteile im Vergleich zu anderen Bildmerkmalen. Zum einen liefern sie eine sparse Repräsentation von Bildern, d.h. das Resultat der Shearlet Transformation hat nur an Pixeln, die einem Kantenpunkt entsprechen, einen deutlichen Betrag. Zum anderen können durch das Multiskalen-Framework der Shearlets Objektstrukturen direkt auf verschiedenen Skalen untersucht werden. Wir zeigen, dass unsere Shearlet Bildmerkmale die besten Ergebnisse mit handgefertigten Bildmerkmalen im Caltech Benchmark liefern.

Aktuell verwenden die besten Verfahren im Caltech Benchmark CNN Modelle. Daher analysieren wir die Fähigkeit von Shearlets, CNNs weiter zu verbessern. Wir integrieren Shearlet Filter in dem ersten Layer eines CNN, da einige gelernte Filter dieses Layers eine Ähnlichkeit zu gerichteten Filtern zur Kantendetektion aufweisen [72]. Wir zeigen, dass wir mit der Shearlet Integration die Klassifikations- und Detektionsergebnisse verbessern können im Vergleich zu den originalen Netzen, wenn wir diese auf denselben Daten trainieren. Wir stellen zudem fest, dass ein Training auf umfangreichen Datensätzen, wie z.B. ImageNet [24], benötigt wird um die Detektionsergebnisse von führenden Algorithmen zu erreichen. Wir erachten diese Feststellung als Indiz für die enorme Macht von Daten für Deep Learning Algorithmen.

Eine der Hauptanwendungen der Fußgängererkennung liegt im Automotive Bereich. Die Verwendung in aktuellen Kollisionswarn- und Eingriffsystemen [40] als auch in zukünftigen Systemen autonomen Fahrens benötigt Algorithmen, welche auf eingebetteten Systemen lauffähig sind. Aus diesem Grund portieren wir unseren grundlegenden Algorithmus auf ein eingebettetes System mit marktfähigem Preis. Durch ein sorgsames Software Design und Laufzeitoptimierung sind wir in der Lage unseren Algorithmus mit einer praktisch sinnvollen Frame Rate von 10 fps zu betreiben.

Contents

"The secret of getting ahead is getting started."
Mark Twain

1

Introduction

The detection of pedestrians is currently a key problem in the area of computer vision. One main reason for it is the diversity of practical applications using pedestrian detection. For example, it is used in *Advanced Driver Assistance Systems (ADAS)* to prevent an imminent collision of the car with a pedestrian by initiating emergency braking. In 2016 the committee of the Euro NCAP introduced an evaluation of AEB Pedestrian systems *(AEB = Autonomous Emergency Braking)* for the safety assessment of passenger cars. According to the Euro NCAP 2020 road map [37], the importance of AEB systems preventing accidents with vulnerable road users such as pedestrians will be further increased in the overall assessment. According to the German Federal Statistical Office [114], pedestrians are the weakest road users. As shown in Figure 1.1, they make up 7.9% of injured in total and 15.5% of fatally injured persons for road accidents in Germany, 2015.

In addition to AEB systems, which are already included in modern road legal cars, the development of *autonomous driving* vehicles strengthens the need of efficient and powerful algorithms for pedestrian detection. Figure 1.2 shows the analysis of emerging technologies according to Gartner [39] in 2017. The expectation concerning autonomous driving is currently located at a peak point. Since autonomous vehicles are enabled by Machine Learning and Deep Learning, these concepts are rated similarly.

The popularity of autonomous driving, Machine Learning and Deep Learning enabled the development of a huge amount of pedestrian detection algorithms in the recent years. In classical pedestrian detection algorithms, a machine learning method, also called *classifier,* decides if an image contains a pedestrian or not based on the measured feature values. According to Benenson et al. [8], different types of classifiers are commonly used, e.g. Support Vector Machines (SVM) and AdaBoost, whereas no classifier type has shown to be better suited for pedestrian detection than another. Therefore, a main focus is on the informative content of the image features. The more meaningful the features, the higher is the quality of the detection algorithm.

In the past, many different kinds of features have been proposed for pedestrian detection. For an overview, see [8, 31, 129]. As a major breakthrough, Dalal and Triggs [22] established the so-called *Histogram of Oriented Gradients (HOG)* features. Here, the image is first divided into spatial cells and a histogram of gradient directions is built over the pixels of the corresponding

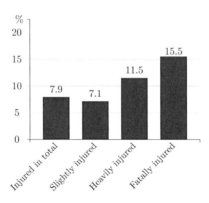

Figure 1.1: Proportion of pedestrians for road accidents in Germany, 2015 [114].

cell. Finally, the local histograms are accumulated and normalized over a block scheme. The *Aggregated Channel Features (ACF)* detector by Dollár et al. [27, 28] also uses gradient histograms with a different computation method and in addition gradient magnitude as features, yielding better evaluation results than the original HOG. The source code of the ACF detector is available as part of the PMT toolbox [26]. As shown by Benenson et al. [8] in 2014, all best performing pedestrian detection algorithms to that time used hand-crafted features based on HOG or "HOG-like" features, which may encode richer information from the original feature data [7, 96, 99, 105, 128, 130].

Since then, several approaches utilizing *Convolutional Neural Network (CNN)* models arose [9, 11, 12, 32, 84, 101, 117, 127]. In CNN models, features are not hand-crafted any longer but learned during the training process. Current results on the Caltech Pedestrian Detection Benchmark [30] show advantages of approaches using CNN models.[1] Several CNN algorithms still use gradient features of the ACF detector [27, 28]. Either in combination with learned features [12] or during object proposal generation for the final classification using a CNN model [84]. Ohn-Bar and Trivedi [99] also used gradient features of the ACF [27, 28] and LDCF [96] detectors during their study of the modeling limitations of boosted decision tree classifiers. They analyzed the impact of the modeling capacity of weak learners, data set size and data set properties. With the employment of their findings, Ohn-Bar and Trivedi achieved the best known results in the Caltech benchmark among non-CNN algorithms. However, the authors did not analyze the impact of alternative image features.

The aim of this thesis is to investigate the applicability of shearlets to improve both currently prevalent algorithm types for pedestrian detection, classical detection algorithms using hand-crafted feature detectors as well as approaches using CNNs. We have chosen the shearlet framework since it guarantees a unified treatment of the continuous as well as the discrete setting [74]. According to the theory on shearlets, they provide optimally sparse approximations of certain image models [52, 76]. Furthermore, they allow the characterization of edges in images [53, 78]. On the practical side, shearlets have been applied in different image processing tasks, e.g. denoising [35, 86] and edge detection [125]. Hereby, we consider our intention to merge theory and practical application as promising.

---

[1]Benchmark results are available and updated frequently at http://www.vision.caltech.edu/Image_Datasets/ CaltechPedestrians.

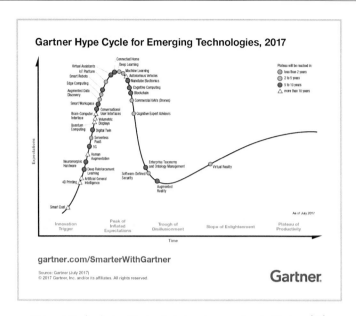

Figure 1.2: Analysis of technology trends according to Gartner [39].

Furthermore, the multiscale framework of shearlets is an extension of *wavelets*, which have been already used by Papageorgiou and Poggio [106] at the early stages of the research on object detection. Viola and Jones [118] built up on this approach and used filters reminiscent to Haar wavelet basic functions for the development of their pioneer object detection algorithm. Wavelets were introduced by Goupillaud, Grossmann and Morlet [44, 49] in order to analyze one-dimensional signals. Wavelet systems consist of analyzing functions which are dilated and translated versions of a mother function $\psi \in L^2(\mathbb{R})$, i.e.

$$\psi_{a,b}(x) = \frac{1}{\sqrt{a}} \psi \left(\frac{x-b}{a} \right), \quad \text{for } a > 0, \ b \in \mathbb{R}.$$

As Kutyniok and Labate [75] describe, wavelets have disadvantages in processing multivariate data such as images. The two dimensional extension of wavelets using a tensor product is able to optimally capture pointwise singularities. Yet this wavelet extension is not able to accurately deal with well distributed singularities such as curved edges. The main reason is that each component of the tensor is a one dimensional function, isotropically dilated from a mother function. To overcome this disadvantage, several approaches of incorporating a *directional sensitivity* have been established [3, 6, 67, 68, 112]. However, these approaches do not actually constitute an extension of the wavelet framework to the multidimensional case. As a matter of fact, they do not yield to an optimal approximation of multivariate data containing well distributed singularities.

Candès and Donoho [13, 14, 15] first provided with *curvelets* a system which is able to optimally approximate bivariate functions with anisotropic features. Just as wavelets, curvelets also set up a sequence of functions which are given on different scales and locations. In addition, these functions are rotated to several orientations in order to incorporate directionality. The utilization of rotations is one main drawback of curvelets since they do not comply with the Cartesian grid.

Correspondingly, a direct numerical implementation of this approach is not possible. As a result, Do and Vetterli [25] introduced *contourlets* as a pure discrete version of curvelets. However, this approach does not provide a proper theoretical consideration of the continuous setting.

Guo, Kutyniok, Labate, Lim and Weiss [51, 81] first introduced shearlets, which build up on *wavelets with composite dilations* [55, 56, 57]. The name of this framework refers to the utilization of shearing to steer the analyzing function to the desired direction. This fact overcomes the disadvantage of curvelets since the shearing operation conforms to the Cartesian grid. Therefore, shearlets provide a direct transition from the continuous to the discrete setting and direct implementations. These are very desirable properties for a theoretical as well as a practical treatment and application.

Due to the incorporation of a directional sensitivity, the shearlet transform of an image allows to capture the directional information of edges. Similar information is provided by gradient histograms. The consideration of shearlets has several advantages. At first, they provide a sparse representation of the image which means that the shearlet transform only has magnitude with considerably high absolute value at pixels that correspond to edge points. Second, due to the multi-scale framework of shearlets the structure of objects can directly be investigated at different scales. Schwartz et al. [111] introduced a simple hand-crafted image feature based on the shearlet transform and applied it to texture classification and face identification. In our work concerning hand-crafted features, we define more complex feature types based on the shearlet transform. Furthermore, we design shearlets dedicated to precisely locate edge structures in images, since this is crucial for the detection quality using our features. Finally, we integrate these shearlets in CNNs in order to improve the detection performance of state-of-the-art CNN algorithms for pedestrian detection.

Contributions

In the following, we present our contribution to the development of pedestrian detection algorithms and the deployment of the shearlet transform to image analysis tasks.

Compactly Supported Shearlets for Image Analysis

In our work, we analyze different types of shearlets and their digital realizations concerning their ability for tasks of image analysis such as edge detection. A precise location of structures in images is of main importance if one is interested in the definition of image features giving structural information. Not surprisingly, shearlets with compact support in time domain provide a better edge localization than band-limited shearlets. We define a new type of compactly supported shearlets, called local precision shearlets, which shows an improved edge detection quality in comparison to other implementations. Furthermore, we prove that the corresponding local precision shearlet transform forms a frame.

Using Shearlet Features for Pedestrian Detection

Based on the local precision shearlet transform we define different types of image features. Since the shearlet transform can be used to extract directional information about structures in images the resulting shearlet features can be seen as appearance features. In other words shearlet features can provide information about the appearance of objects in images. Currently, the dominant appearance features used in nearly all pedestrian detection algorithms are based on image gradients. Here, the directional information of image structures is represented by the gradient direction. Since shearlet features are providing similar information, they can be seen as an alternative to gradient features. We analyze the impact of replacing gradient by shearlet features. In fact, we show that our shearlet features outperform gradient features in the application of pedestrian detection. Furthermore, we set up a filterbank for an intermediate filtering layer between image feature computation and classification with our shearlets. The application of this filterbank yields the currently best performing hand-crafted feature detector in the Caltech Pedestrian Detection Benchmark [30].

Integration of Shearlets in Deep Learning Methods

With hand-crafted shearlet features as foundation we exploit the possibilities to integrate our shearlets in Deep Learning methods such as CNNs. We aim to use shearlet filters at early convolutional layers of a CNN instead of learned ones in order to improve its detection results. The idea is that early CNN layers intuitively perform an edge detection, whereas shearlets provide optimal filters for this task. Moreover, they can provide a good base for learning filters of deeper layers. We find that the integration of shearlets in CNNs improves the results on pedestrian classification and detection compared to corresponding reference networks trained on the same data.

Embedded Realization

Besides the considerations for improving the detection quality, we focus on an embedded realization of the shearlet transform. The computation of the shearlet coefficients is the core component concerning runtime consumption in our hand-crafted feature detection algorithms. Therefore, runtime optimization is crucial if one is interested in an embedded application using shearlets. To our knowledge, our implementation of the shearlet transform is the first one considering an application on an embedded target. Actually, our optimized realization is able to run our base pedestrian detection algorithm on an embedded target with an adequate frame rate.

Outline

The thesis is organized as follows. Chapter 2 reviews the work on shearlets with a special focus on image analysis. In particular, we analyze current realizations of the shearlet transform and their applicability to be used as a basis for image feature computation. In Chapter 3, we define our own type of shearlets specially designed for image analysis and show their practical advantages. Moreover, we analyze their theoretical properties especially in the context of frame theory. The theoretical analysis concerning the applicability for edge detection using our shearlets is done in Chapter 4. We then use our shearlets in Chapter 5 to define hand-crafted image features for the task of pedestrian detection. Furthermore, we set up two detection algorithms. One of them is using basic shearlet features and the other one is applying an intermediate filtering layer with a shearlet filterbank. Chapter 6 analyzes how our shearlets can be used to improve the quality of pedestrian detection algorithms utilizing CNNs. Besides all considerations concerning improvement of detection quality, Chapter 7 deals with embedded realization of the shearlet transform and our base detection algorithm.

Review on Shearlets for Image Analysis

The shearlet framework is a variation of the wavelet scheme for multidimensional data which incorporates a directional sensitivity to capture anisotropic information. For an introduction and details on wavelets see [90]. Like wavelets, shearlets are set up given a single or finite set of generating functions, also called *mother shearlets*, which are translated along the signal and dilated. To control the directional selectivity, the mother shearlet is sheared. The final translated, dilated and sheared function is defined by

$$\psi_{a,s,t} := \det\left(A_a\right)^{-\frac{1}{2}} \psi\left(A_a^{-1} S_s^{-1}\left(\cdot - t\right)\right), \tag{2.1}$$

with scaling matrix A_a and shearing matrix S_s given by

$$A_a := \begin{pmatrix} a & 0 \\ 0 & \sqrt{a} \end{pmatrix}, \ a > 0$$

and

$$S_s := \begin{pmatrix} 1 & s \\ 0 & 1 \end{pmatrix}, \ s \in \mathbb{R}.$$

As described in [75], the more general matrices

$$A_{a,\alpha} := \begin{pmatrix} a & 0 \\ 0 & a^\alpha \end{pmatrix}$$

can be used instead of A_a where the parameter $\alpha \in (0,1)$ controls the *degree of anisotropy*. An extensive study of transforms based on scalings with $A_{a,\alpha}$ can be found in [48].

One major goal in the analysis of signals is to define a representation system of functions using sets of $\psi_{a,s,t}$. To guarantee stability of the representation while allowing nonunique decompositions, the concept of a *frame* has been established [23, 33]. A sequence $(\varphi_i)_{i\in I}$ in a Hilbert space $\mathcal{H}$ is called a frame for $\mathcal{H}$, if there exist constants $0 < A \leq B < \infty$ such that

$$A \left\|x\right\|^2 \leq \sum_{i\in I} |\langle x, \varphi_i\rangle|^2 \leq B \left\|x\right\|^2 \quad \text{for all } x \in \mathcal{H}.$$

The frame is called *tight* if the frame constants A and B fulfill $A = B$. If $A = B = 1$ is fulfilled, the frame is called a *Parseval frame*. As described by Christensen [17], in case that $(\varphi_i)_{i \in I}$ is a frame, a signal $x \in \mathcal{H}$ can be reconstructed from its frame coefficients by the formula

$$x = \sum_{i \in I} \langle x, \varphi_i \rangle \, S^{-1} \varphi_i, \tag{2.2}$$

where $S \colon \mathcal{H} \to \mathcal{H}$ is called the *frame operator* and given by

$$S\,(x) = \sum_{i \in I} \langle x, \varphi_i \rangle \, \varphi_i.$$

If $(\varphi_i)_{i \in I}$ is a Parseval frame we have $S = I_{\mathcal{H}}$, with the identity operator $I_{\mathcal{H}}$ on $\mathcal{H}$. Consequently, this fact leads to the reconstruction formula

$$x = \sum_{i \in I} \langle x, \varphi_i \rangle \, \varphi_i. \tag{2.3}$$

In the following sections we will provide theoretical key results about shearlets on the continuous as well as the discrete setting. Furthermore, we will provide an overview over current implementations of the discrete shearlet transform and their properties concerning image analysis.

2.1   Continuous Shearlet Systems

We first give the definition of shearlet systems and the corresponding transform in the continuous setting, i.e. for $(a, s, t) \in \mathbb{R}^+ \times \mathbb{R} \times \mathbb{R}^2$. This introduction is based on the description of Kutyniok and Labate [75]. For implementation purposes, the parameters have to be defined on a discrete subset of $\mathbb{R}^+ \times \mathbb{R} \times \mathbb{R}^2$, which will be described in the subsequent section.

Definition 2.1. For $\psi \in L^2(\mathbb{R}^2)$, the continuous shearlet system $SH\,(\psi)$ is defined by

$$SH\,(\psi) := \left\{ \psi_{a,s,t} = \det\,(A_a)^{-\frac{1}{2}} \, \psi \left(A_a^{-1} S_s^{-1} \, (\cdot - t) \right) : a > 0, s \in \mathbb{R}, t \in \mathbb{R}^2 \right\}.$$

Just like the continuous wavelet transform [90], the continuous shearlet transform of a function $f \in L^2(\mathbb{R}^2)$ is defined as the mapping from f to the coefficients of f associated to the shearlet $\psi_{a,s,t}$.

Definition 2.2. The *continuous shearlet transform* of a signal $f \in L^2(\mathbb{R}^2)$ for $(a, s, t) \in \mathbb{R}^+ \times \mathbb{R} \times \mathbb{R}^2$ is defined by

$$
\begin{aligned}
\mathcal{SH}_\psi\,(f)\,(a, s, t) &:= \langle f, \psi_{a,s,t} \rangle \\
&= \int_{\mathbb{R}^2} f\,(x)\,\overline{\psi_{a,s,t}\,(x)}\mathrm{d}x.
\end{aligned}
$$

The resulting values $\mathcal{SH}_\psi\,(f)\,(a, s, t) = \langle f, \psi_{a,s,t} \rangle$ are also called *shearlet coefficients*. As shown in [54, 125], the shearlet transform of an image f characterizes the location and the orientation of edges. The characterization is given via the asymptotic behavior of the shearlet transform at fine scales. More precisely for an image point t not being an edge, $|\mathcal{SH}_\psi\,(f)\,(a, s, t)|$ decays rapidly for $a \to 0$ for each $s \in \mathbb{R}$. $|\mathcal{SH}_\psi\,(f)\,(a, s, t)|$ is said to *decay rapidly* if for any $N \in \mathbb{N}$, there is a $c_N > 0$ such that $|\mathcal{SH}_\psi\,(f)\,(a, s, t)| \leq c_N a^N$, as $a \to 0$. For an edge point t, $|\mathcal{SH}_\psi\,(f)\,(a, s, t)|$ decays rapidly for $a \to 0$ unless s equals the normal orientation to the edge at point t. Then one has $\left| \mathcal{SH}_\psi \left(f \right) \left(a, s, t \right) \right| \sim a^{\frac{3}{4}}$. More details on the theoretical analysis of edge detection using shearlets are given in Section 2.4. For a more intuitive understanding one can say that high absolute values of shearlet coefficients only appear at edge points and if the orientation of the shearlet and the edge correspond to each other. Moreover, the finer the scale of the shearlet, the finer edges can be detected. This ability will be used in Section 5.2 to define meaningful image features for the task of pedestrian detection.

As described by Kutyniok and Labate [75], the mother function ψ is usually defined by its Fourier transform $\hat{\psi}$ to give a factorization of the frequency domain by defining

$$\hat{\psi} \left(\omega_1, \omega_2 \right) := \hat{\psi}_1 \left(\omega_1 \right) \hat{\psi}_2 \left(\frac{\omega_2}{\omega_1} \right).$$

The components $\hat{\psi}_1$ and $\hat{\psi}_2$ are commonly chosen to be compactly supported functions fulfilling the following *admissibility* condition. This condition on ψ is of high importance since it leads to the result that the associated shearlet transform is an isometry and to the existence of a reconstruction formula.

Definition 2.3. If $\psi \in L^2 \left(\mathbb{R}^2 \right)$ fulfills the condition

$$\int_{\mathbb{R}^2} \frac{\left| \hat{\psi} \left(\xi_1, \xi_2 \right) \right|^2}{\xi_1^2} \mathrm{d}\xi_2 \mathrm{d}\xi_1 < \infty, \tag{2.4}$$

then it is called an *admissible shearlet*.

Further desirable properties of the shearlet mother function ψ in the frequency domain are given as follows.

Definition 2.4 ([46, 47]). For $p \in \mathbb{N}$, a function $\psi \in L^2(\mathbb{R}^2)$ possesses p (directional) vanishing moments in x_1 direction if

$$\int_{\mathbb{R}^2} \frac{\left| \hat{\psi} \left(\xi_1, \xi_2 \right) \right|^2}{\left| \xi_1 \right|^{2p}} \mathrm{d}\xi_2 \mathrm{d}\xi_1 < \infty. \tag{2.5}$$

A function $f \in L^2(\mathbb{R}^2)$ has Fourier decay of order l_i in the i-th variable if $\left| \hat{f} \left(\xi \right) \right| \lesssim \left| \xi_i \right|^{l_i}$.

Remark 2.5 ([46]). The rationale for the denotation of vanishing moments in the previous definition is that condition 2.5 is (almost) equivalent to

$$\int_{\mathbb{R}} \psi \left(x_1, x_2 \right) x_1^l \mathrm{d}x_1 = 0, \quad \text{for all } x_2 \in \mathbb{R}^2, \ l < p,$$

provided that ψ has sufficient spatial decay.

For the sake of completeness, we state the definition of vanishing moments in the one-dimensional case.

Definition 2.6 ([90]). For $p \in \mathbb{N}$, a function $\psi \colon \mathbb{R} \to \mathbb{R}$ has p vanishing moments if

$$\int_{\mathbb{R}} \psi \left(x \right) x^l \mathrm{d}x = 0,$$

for all $l \in \mathbb{N}$ and $l < p$.

Theorem 2.7 ([21]). *Given an admissible $\psi \in L^2(\mathbb{R}^2)$, define*

$$c_\psi^- := \int_\mathbb{R} \int_{-\infty}^0 \frac{\left|\hat{\psi}\left(\xi_1, \xi_2\right)\right|^2}{\xi_1^2} \mathrm{d}\xi_2 \mathrm{d}\xi_1 \quad and \quad c_\psi^+ := \int_\mathbb{R} \int_0^\infty \frac{\left|\hat{\psi}\left(\xi_1, \xi_2\right)\right|^2}{\xi_1^2} \mathrm{d}\xi_2 \mathrm{d}\xi_1.$$

If $c_\psi^- = c_\psi^+ = c_\psi$, then the shearlet transform is a c_ψ-multiple of an isometry.

For a signal $f \in L^2(\mathbb{R}^2)$, we have the following reconstruction formula. Here, we use the notion of an *approximate identity*, which is defined in Appendix A.

Theorem 2.8 ([21]). *Let $\psi \in L^2(\mathbb{R}^2)$ be an admissible shearlet with $c_\psi^+ = c_\psi^- = 1$. Furthermore, let $(\rho_n)_{n=1}^\infty$ be an approximate identity such that $\rho_n \in L^2(\mathbb{R}^2)$ and $\rho_n(x) = \rho_n(-x)$ for all $x \in \mathbb{R}^2$. For all $f \in L^2(\mathbb{R}^2)$, we have $\lim_{n\to\infty} \|f - f_n\|_2 = 0$ with*

$$f_n(x) = \int_{\mathbb{R}^2} \int_\mathbb{R} \int_{\mathbb{R}^+} \mathcal{SH}_\psi(f)(a, s, t)(\rho_n * \psi_{a,s,t})(x) a^{-3} \mathrm{d}a \, \mathrm{d}s \, \mathrm{d}t.$$

As described by Kutyniok and Labate [75] an important example for the mother function ψ is given by the *classical shearlet* [51, 55, 81].

Definition 2.9. Let $\psi \in L^2(\mathbb{R}^2)$ be given by

$$\hat{\psi}(\omega_1, \omega_2) = \hat{\psi}_1(\omega_1)\hat{\psi}_2\left(\frac{\omega_2}{\omega_1}\right), \tag{2.6}$$

where $\psi_1 \in L^2(\mathbb{R})$ is a discrete wavelet in the sense that it satisfies the discrete Calderón condition, given by

$$\sum_{j\in\mathbb{Z}}\left|\hat{\psi}_1\left(2^{-j}\omega\right)\right|^2 = 1$$

for a.e. $\omega \in \mathbb{R}$, with $\hat{\psi}_1 \in C^\infty(\mathbb{R})$ and supp $\hat{\psi}_1 \subseteq \left[-\frac{1}{2}, -\frac{1}{16}\right] \cup \left[\frac{1}{16}, \frac{1}{2}\right]$. Furthermore, $\psi_2 \in L^2(\mathbb{R})$ is a function that fulfills

$$\sum_{k=-1}^1 \left|\hat{\psi}_2(\omega + k)\right|^2 = 1$$

for a.e. $\omega \in [-1, 1]$, satisfying $\hat{\psi}_2 \in C^\infty(\mathbb{R})$ and supp $\hat{\psi}_2 \subseteq [-1, 1]$. Then ψ is called a *classical shearlet*.

2.2   Discrete Shearlet Systems

In order to be able to digitally implement the shearlet transform, the shearlet parameters have to be sampled such that a discrete set is made up. Discrete shearlet systems are defined by taking only those shearlets that are associated with a discrete subset of $\mathbb{R}^+ \times \mathbb{R} \times \mathbb{R}^2$. In the following, we will mainly use the notation of [73].

Definition 2.10. Let $\psi \in L^2(\mathbb{R}^2)$ and $\Gamma_I \subseteq \mathbb{S}$ given by

$$\Gamma_I := \left\{\left(a_j, s_{j,k}, S_{s_{j,k}} A_{a_j} cm\right) : j, k \in \mathbb{Z}, m \in \mathbb{Z}^2\right\}, \tag{2.7}$$

with $a_j \in \mathbb{R}^+$, $s_{j,k} \in \mathbb{R}$ and sampling constant $c > 0$. An *irregular discrete shearlet system* associated with ψ and Γ_I, denoted by $SH(\psi, \Gamma_I)$, is defined by

$$SH\left(\psi, \Gamma_I\right) := \left\{ \psi_{j,k,m} = \det\left(A_{a_j}\right)^{-\frac{1}{2}} \psi\left(S_{s_{j,k}} A_{a_j} \cdot -cm\right) : j, k \in \mathbb{Z}, \, m \in \mathbb{Z}^2 \right\}.$$

The shearlet system $SH(\psi, \Gamma_R)$ given by

$$SH\left(\psi, \Gamma_R\right) := \left\{ \psi_{j,k,m} = a^{-\frac{3}{4}j} \psi\left(S_{bka^{j/2}} A_{a^j} \cdot -cm\right) : j, k \in \mathbb{Z}, \, m \in \mathbb{Z}^2 \right\}$$

with $a > 1$, b, $c > 0$ and the special parameter set $\Gamma_R \subseteq \mathbb{S}$ defined by

$$\Gamma_R := \left\{ \left(a^j, bka^{\frac{j}{2}}, S_{bka^{j/2}} A_{a^j} cm\right) : j, k \in \mathbb{Z}, \, m \in \mathbb{Z}^2 \right\} \tag{2.8}$$

is called a *regular discrete shearlet system*.

The *discrete shearlet transform* is defined in analogy to the continuous case.

Definition 2.11. For $\psi \in L^2(\mathbb{R}^2)$, the *discrete shearlet transform* of a signal $f \in L^2(\mathbb{R}^2)$ is defined by

$$\mathcal{SH}_\psi\left(f\right)(j, k, m) := \langle f, \psi_{j,k,m} \rangle$$

with $\psi_{j,k,m} \in \{\Gamma_I, \Gamma_R\}$.

The following theorem provides sufficient conditions under which an irregular shearlet system is a frame for $L^2(\mathbb{R}^2)$.

Theorem 2.12 ([73]). *Let $c > 0$ be fixed and, for j, $k \in \mathbb{Z}$, let $a_j \in \mathbb{R}^+$ and $s_{j,k} \in \mathbb{R}$. Define $\Gamma_I \subseteq \mathbb{S}$ as in (2.7). Further, let $\psi \in L^2(\mathbb{R}^2)$ and set*

$$\varphi\left(n, SH\left(\psi, \Gamma_I\right)\right) := \operatorname*{ess\,sup}_{\xi \in \mathbb{R}^2} \sum_{j,k \in \mathbb{Z}} \left| \hat{\psi}\left(A_{a_j} S_{s_{j,k}}^T \xi\right) \right| \left| \hat{\psi}\left(A_{a_j} S_{s_{j,k}}^T \xi + n\right) \right| \quad \text{for a.e. } n \in \mathbb{R}^2 \tag{2.9}$$

as well as

$$\tilde{\varphi}\left(SH\left(\psi, \Gamma_I\right)\right) := \sum_{j,k \in \mathbb{Z}} \left| \hat{\psi}\left(A_{a_j} S_{s_{j,k}}^T \xi\right) \right|^2. \tag{2.10}$$

If there exist $0 < C \leq D < \infty$ such that

$$C \leq \tilde{\varphi}\left(SH\left(\psi, \Gamma_I\right)\right) \leq D \quad \text{for a.e. } \xi \in \mathbb{R}^2 \tag{2.11}$$

and

$$\sum_{n \in \mathbb{Z}^2, n \neq 0} \sqrt{\varphi\left(\frac{1}{c}n, SH\left(\psi, \Gamma_I\right)\right) \varphi\left(-\frac{1}{c}n, SH\left(\psi, \Gamma_I\right)\right)} =: E < C, \tag{2.12}$$

then $SH(\psi, \Gamma)$ is a frame for $L^2(\mathbb{R}^2)$ with frame bounds A, B satisfying

$$\frac{1}{c^2}\left(C - E\right) \leq A \leq B \leq \frac{1}{c^2}\left(D + E\right).$$

From this result, Lim [86] derived sufficient conditions for separable shearlets to provide a frame for $L^2(\mathbb{R}^2)$.

Theorem 2.13 ([86]). *Let Γ_R be given by (2.8) with $a > 1$ and $b = 1$ and scaling matrix $A_{a,\alpha}$, $\alpha \in [1/2, 1)$. We set $\beta > 0$ and $\gamma > 2(\beta + 2)$ and assume that $\beta' \geq \beta + \gamma$ and $\gamma' \geq \beta' - \beta + \gamma$. We define $\psi(x_1, x_2) = \psi_1(x_1)\psi_2(x_2)$ such that*

$$\left| \hat{\psi}_1(\xi_1) \right| \leq K_1 \frac{|\xi_1|^{\beta'}}{\left(1 + |\xi_1|^2 \right)^{\gamma'/2}} \tag{2.13}$$

and

$$\left| \hat{\psi}_2(\xi_2) \right| \leq K_2 \left(1 + |\xi_2|^2 \right)^{-\gamma'/2} \tag{2.14}$$

with $K_1, K_2 > 0$. If

$$\operatorname*{ess\,inf}_{|\xi_2| \leq 1/2} \left| \hat{\psi}_2(\xi_2) \right|^2 \geq K_3 > 0 \tag{2.15}$$

and

$$\operatorname*{ess\,inf}_{\zeta a^{-1} \leq |\xi_1| \leq \zeta} \left| \hat{\psi}_1(\xi_1) \right|^2 \geq K_4 > 0 \quad \text{for } 0 < \zeta \leq \min\left(1, \frac{a}{2} \right) \tag{2.16}$$

then there exists $c_0 > 0$ such that the regular shearlet system $SH(\psi, \Gamma_R)$ is a frame for $L^2(\mathbb{R}^2)$ for all $c \leq c_0$.

According to Lim [86], the assumptions of Theorem 2.13 imply that ψ has sufficient vanishing moments and fast decay in the frequency domain. If they are fulfilled the regular shearlet system based on separable shearlets provides a frame for $L^2(\mathbb{R}^2)$. As we will see later, this result transfers to cone-adapted shearlet systems, which are described in the next section.

2.3   Cone-adapted Shearlet Systems

The shearlet systems described above do have a directional bias [75]. For example, for detecting an edge along the x_1-axis of an image, $s \to \infty$ would be necessary. Obviously, this is a drawback for practical applications. According to Kutyniok and Labate [75], a way to resolve it is to define *cone-adapted shearlets*, tiling the frequency domain in a horizontal and a vertical cone and defining shearlets separately for each cone. We set up a low-frequency region $\mathcal{R} := \{(\xi_1, \xi_2) \colon |\xi_1|, |\xi_2| \leq 1/2\}$, horizontal cones $\mathcal{C}_1 \cup \mathcal{C}_3 := \{(\xi_1, \xi_2) \colon |\xi_2/\xi_1| \leq 1, |\xi_1| > 1/2\}$ and vertical cones $\mathcal{C}_2 \cup \mathcal{C}_4 := \{(\xi_1, \xi_2) \colon |\xi_2/\xi_1| > 1, |\xi_2| > 1/2\}$. In this section, we restrict the description to the case of discrete shearlet systems.

Definition 2.14. Let $\phi, \psi, \tilde{\psi} \in L^2(\mathbb{R}^2)$ and $\Gamma, \tilde{\Gamma} \subseteq \mathbb{S}$ given by

$$\Gamma := \left\{ \left(a_j, s_{j,k}, S_{s_{j,k}} A_{a_j} cm \right) : j \in J, \, k \in K, \, m \in \mathbb{Z}^2 \right\}, \tag{2.17}$$

and

$$\tilde{\Gamma} := \left\{ \left(\tilde{a}_j, \tilde{s}_{j,k}, S_{\tilde{s}_{j,k}}^T \tilde{A}_{\tilde{a}_j} cm \right) : j \in \tilde{J}, \, k \in \tilde{K}, \, m \in \mathbb{Z}^2 \right\}, \tag{2.18}$$

with $a_j \in \mathbb{R}^+$, $s_{j,k} \in \mathbb{R}$ for $j \in J \subseteq \mathbb{Z}$, $k \in K \subseteq \mathbb{Z}$, $\tilde{a}_j \in \mathbb{R}^+$, $\tilde{s}_{j,k} \in \mathbb{R}$ for $j \in \tilde{J} \subseteq \mathbb{Z}$, $k \in \tilde{K} \subseteq \mathbb{Z}$, $\tilde{A}_{\tilde{a}_j} = \operatorname{diag}(\sqrt{\tilde{a}_j}, \tilde{a}_j)$ and sampling constants $c > 0$. An *irregular cone-adapted discrete shearlet system* $SH(\phi, \psi, \tilde{\psi}, \Gamma, \tilde{\Gamma})$, is defined by

$$SH\left(\phi, \psi, \tilde{\psi}, \Gamma, \tilde{\Gamma} \right) := \Phi\left(\phi, c \right) \cup \Psi\left(\psi, \Gamma \right) \cup \tilde{\Psi}\left(\tilde{\psi}, \tilde{\Gamma} \right)$$

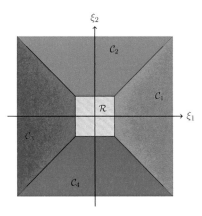

Figure 2.1: Illustration of tiling the frequency plane into cones $\mathcal{C}_i$, $i = 1, \ldots, 4$.

where

$$\Phi\left(\phi, c\right) \quad := \quad \left\{\phi_m = \phi\left(\cdot - cm\right) : m \in \mathbb{Z}^2\right\},$$

$$\Psi\left(\psi, \Gamma\right) \quad := \quad \left\{\psi_{j,k,m} = a_j^{-\frac{3}{4}}\psi\left(S_{s_{j,k}}A_{a_j} \cdot -cm\right) : j \in J,\, k \in K,\, m \in \mathbb{Z}^2\right\},$$

$$\tilde{\Psi}\left(\tilde{\psi}, \tilde{\Gamma}\right) \quad := \quad \left\{\tilde{\psi}_{j,k,m} = \tilde{a}_j^{-\frac{3}{4}}\tilde{\psi}\left(S_{\tilde{s}_{j,k}}^T\tilde{A}_{\tilde{a}_j} \cdot -cm\right) : j \in \tilde{J},\, k \in \tilde{K},\, m \in \mathbb{Z}^2\right\}.$$

Usually the functions are chosen such that ϕ is associated with the low-frequency region $\mathcal{R}$, ψ with $\mathcal{C}_1 \cup \mathcal{C}_3$ and $\tilde{\psi}$ with $\mathcal{C}_2 \cup \mathcal{C}_4$. Similar to the previous section, the regular variant of a cone-adapted shearlet system reads as follows.

Definition 2.15. Let ϕ, ψ, $\tilde{\psi} \in L^2(\mathbb{R}^2)$, $a > 1$ and b, $c > 0$. The *regular cone-adapted discrete shearlet system* $SH(\phi, \psi, \tilde{\psi}, c)$, is defined by

$$SH\left(\phi, \psi, \tilde{\psi}, c\right) := \Phi\left(\phi, c\right) \cup \Psi\left(\psi, c\right) \cup \tilde{\Psi}\left(\tilde{\psi}, c\right)$$

where

$$\Phi\left(\phi, c\right) := \left\{\phi_m = \phi\left(\cdot - cm\right) : m \in \mathbb{Z}^2\right\},$$

$$\Psi\left(\psi, c\right) := \left\{\psi_{j,k,m} = a^{-\frac{3}{4}j}\psi\left(S_{bka^{j/2}}A_{a^j} \cdot -cm\right) : j \geq 0, |k| \leq \left\lceil a^{j/2}\right\rceil, m \in \mathbb{Z}^2\right\},$$

$$\tilde{\Psi}\left(\tilde{\psi}, c\right) := \left\{\tilde{\psi}_{j,k,m} = a^{-\frac{3}{4}j}\tilde{\psi}\left(S_{bka^{j/2}}^T\tilde{A}_{a^j} \cdot -cm\right) : j \geq 0, |k| \leq \left\lceil a^{j/2}\right\rceil, m \in \mathbb{Z}^2\right\}.$$

The *discrete cone-adapted shearlet transform* in the regular case is then given by the combination of the discrete shearlet transforms of the shearlet system components.

Definition 2.16. For ψ, $\tilde{\psi}$, $\phi \in L^2(\mathbb{R}^2)$ and $a > 1$ and b, $c > 0$, the *discrete cone-adapted shearlet transform* of an image $f \in L^2(\mathbb{R}^2)$ is given by

$$\mathcal{SH}_{\phi,\psi,\tilde{\psi}}\left(f\right)\left(j, k, m\right) := \left(\langle f, \phi_m\rangle, \langle f, \psi_{j,k,m}\rangle, \left\langle f, \tilde{\psi}_{j,k,m}\right\rangle\right), \tag{2.19}$$

with $j \geq 0$, $|k| \leq \left\lceil a^{j/2}\right\rceil$ and $m \in \mathbb{Z}^2$.

In certain cases, the results for shearlet systems can be transferred to their cone-adapted variants. In particular, Lim [86] derived the following statement.

Corollary 2.17. *Let* $\psi(x_1, x_2) = \psi_1(x_1)\psi_2(x_2)$, $\tilde{\psi}(x_1, x_2) = \psi_1(x_2)\psi_2(x_1)$ *and* $\phi(x_1, x_2) = \psi_2(x_1)\psi_2(x_2)$, *where the functions* ψ_1 *and* ψ_2 *satisfy the assumptions of Theorem 2.13. Then there exists* $c_0 > 0$ *such that* $SH(\phi, \psi, \tilde{\psi}, c)$ *forms a frame of* $L^2(\mathbb{R}^2)$ *for all* $c \leq c_0$.

2.4 Edge Detection using Shearlets

In this section, we will recall major results on the characterization of edge points of a function $f = \chi_R$ with $R \subset \mathbb{R}^2$ based on the properties of its continuous shearlet transform. In other words, the boundary curve of a general region $R \subset \mathbb{R}^2$ will be characterized by the shearlet transform. In the following, we mainly use the notation of [53].

Let $R \subset \mathbb{R}^2$ whose boundary ∂R is a curve of finite length L. Furthermore, let $\vec{\alpha} \colon (0, L) \to \partial R$ be a parametrization of ∂R. For any $t_0 \in (0, L)$ and any $l \geq 0$, we assume that there exist left and right limits $\vec{\alpha}^{(l)}(t_0^-)$ and $\vec{\alpha}^{(l)}(t_0^+)$. With $\vec{n}(t^-)$ and $\vec{n}(t^+)$ we denote the left and right hand side outer normal directions of ∂R at $\vec{\alpha}(t)$. If $\vec{n}(t^-) = \vec{n}(t^+)$, we use $\vec{n}(t)$. Furthermore, we set $\vec{\kappa}(t^-)$, $\vec{\kappa}(t^+)$, or $\vec{\kappa}(t)$, respectively for the curvature of ∂R at $\vec{\alpha}(t)$, i.e. $\vec{\kappa}(t) = \|\vec{\alpha}''(t)\|$. Finally, we say that a shearing parameter s *corresponds to direction* $\vec{n} = \pm(\cos(\theta_0), \sin(\theta_0))$ for $\theta_0 \in [0, 2\pi]$ if $s = \tan(\theta_0)$.

Definition 2.18. *Let* $R \subset \mathbb{R}^2$ *be as described above. A point* $p = \vec{\alpha}(t_0)$ *is a* corner point *of* ∂R *if one of the following conditions holds:*

i. $\vec{\alpha}'\left(t_0^-\right) \neq \vec{\alpha}'\left(t_0^+\right)$

ii. $\vec{\alpha}'\left(t_0^-\right) = \pm\vec{\alpha}'\left(t_0^+\right)$, *but* $\vec{\kappa}\left(t_0^-\right) \neq \vec{\kappa}\left(t_0^+\right)$.

In case of i., we call p a *corner point of first type* and in case of ii., a *corner point of second type*. If $\vec{\alpha}$ is infinitely many times differentiable at t_0, we call $p = \vec{\alpha}(t_0)$ a *regular point* of ∂R.

Guo and Labate [53] showed the following result for the characterization of edge points using classical shearlets.

Theorem 2.19. *Let* $\psi \in L^2(\mathbb{R}^2)$ *be a classical shearlet and* $R \subset \mathbb{R}^2$ *with boundary* ∂R *of length* L *to be smooth except for finitely many corner points.*

i. *If* $p \notin \partial R$, *then for any* $s \in \mathbb{R}$ *we have*

$$\lim_{a \to 0+} a^{-N} \mathcal{SH}_\psi \chi_R(a, s, p) = 0, \quad \text{for all } N > 0.$$

ii. *Let* p *be a regular point of* ∂R.

    (a) *If* $s = s_0$ *does not correspond to the normal direction of* ∂R *at* p, *then*

$$\lim_{a \to 0+} a^{-N} \mathcal{SH}_\psi \chi_R(a, s_0, p) = 0, \quad \text{for all } N > 0.$$

(b) If $s = s_0$ corresponds to the normal direction of ∂R at p, then

$$0 < \lim_{a \to 0+} a^{-\frac{3}{4}} \left| \mathcal{SH}_\psi \chi_R \left(a, s_0, p\right) \right| < \infty.$$

iii. Let $p \in \partial R$ be a corner point.

(a) If p is a corner point of the first type and $s = s_0$ does not correspond to any of the normal directions of ∂R at p, then

$$\lim_{a \to 0+} a^{-\frac{9}{4}} \left| \mathcal{SH}_\psi \chi_R \left(a, s_0, p\right) \right| < \infty.$$

(b) If p is a corner point of the second type and $s = s_0$ does not correspond to any of the normal directions of ∂R at p, then

$$0 < \lim_{a \to 0+} a^{-\frac{9}{4}} \left| \mathcal{SH}_\psi \chi_R \left(a, s_0, p\right) \right| < \infty.$$

(c) If $s = s_0$ corresponds to one of the normal directions of ∂R at p then

$$0 < \lim_{a \to 0+} a^{-\frac{3}{4}} \left| \mathcal{SH}_\psi \chi_R \left(a, s_0, p\right) \right| < \infty.$$

This result shows that for regular points $p \in \partial R$ the shearlet transform using a classical shearlet decays rapidly and asymptotically for $a \to 0$ if s does not correspond to the normal direction at p. If s corresponds to the normal direction, we have

$$\left| \mathcal{SH}_\psi \chi_R \left(a, s, p\right) \right| = \mathcal{O}\left(a^{\frac{3}{4}}\right), \text{ for } a \to 0.$$

The same decay rate is present at a corner point $p \in \partial R$ if s corresponds to a normal direction of ∂R at p. In case s does not correspond to a normal direction of ∂R at a corner point p of the second type, then we have

$$\left| \mathcal{SH}_\psi \chi_R \left(a, s, p\right) \right| = \mathcal{O}\left(a^{\frac{9}{4}}\right), \text{ for } a \to 0.$$

In case $p \in \partial R$ is a corner point of the first type, the decay rate of $\left| \mathcal{SH}_\psi \chi_R \left(a, s, p\right) \right|$ is not slower than $\mathcal{O}(a^{\frac{9}{4}})$ for $a \to 0$ although it might be faster. Figure 2.2 [53] illustrates the decay rates of the shearlet transform for the different categories of points $p \in \mathbb{R}^2$.

Kutyniok and Petersen [78] examined the characterization of edge points using compactly supported shearlets in order to resolve certain issues when using band-limited shearlets. According to the authors, one major issue is that the decay rates of the shearlet transform are not uniform. According to 2.19 a point $p \in \mathbb{R}^2$ is an edge point if

$$\lim_{a \to 0+} a^{-\frac{3}{4}} \left| \mathcal{SH}_\psi \chi_R \left(a, s, p\right) \right| > 0$$

for some $s \in \mathbb{R}$. This limit can be arbitrarily close to 0 and the asymptotic behavior might only be present for very small $a \in \mathbb{R}$. In contrast, Kutyniok and Petersen [78] achieve uniform estimates on the decay rates in a sense that there exist constants $0 < c_1 \le c_2 < \infty$ such that for all $a \in (0, 1)$

$$c_1 a^{\frac{3}{4}} \le \left| \mathcal{SH}_\psi \chi_R \left(a, s, p\right) \right| \le c_2 a^{\frac{3}{4}}$$

for all edge points $p \in \partial R$ and orientations that correspond to the normal direction. Furthermore, another issue of using band-limited shearlets is that both types of corner types show the same decay rates. According to Kutyniok and Petersen[78], corner points of first and second type can be distinguished by different decay rates if one uses compactly supported shearlets. For their findings, a definition of a restricted set of compact sets in $\mathbb{R}^2$ is necessary.

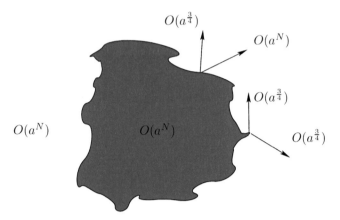

Figure 2.2: Illustration of the decay rates of the shearlet transform using classical shearlets [53].

Definition 2.20. For $\rho > 0$ the set of all sets $R \subset \mathbb{R}^2$ with piecewise smooth boundary with corner points $\{p_i \colon i \in I\}$ and arc-length parametrization $\vec{\alpha}$ such that

   i. $\left\| \vec{\alpha}^{(3)}(t) \right\| \leq \rho$ for all $t \in (0, L)$, $t \notin \vec{\alpha}^{-1}(\{p_i \colon i \in I\})$,

   ii. $\left\| \vec{\alpha}^{(3)}(t^{\pm}) \right\| \leq \rho$ for all $t \in \vec{\alpha}^{-1}(\{p_i\}_{i \in I})$,

will be denoted by $\mathcal{R}_\rho$.

Before the statement of the result of [78], we introduce the denotation of the Sobolev space consisting of all L times weakly differentiable functions $f \in L^2(\mathbb{R}^2)$ with all weak derivatives in $L^2(\mathbb{R}^2)$ by $H^L(\mathbb{R}^2)$. For the basic definition of Sobolev spaces, see Appendix A. Furthermore, we define the ball around $\tilde{s}$ with radius a by $B_a(\tilde{s}) = \{s \in \mathbb{R} \colon |s - \tilde{s}| \leq a\}$ for $a, \tilde{s} \in \mathbb{R}$. With these preparations, the result of [78] is then given by the following theorem. The obtained decay rates of the different types of edge points using compactly supported shearlets are illustrated in Figure 2.3.

Theorem 2.21 ([78]). *Let $R \in \mathcal{R}_\rho$ for $\rho > 0$ and $\psi \in L^2(\mathbb{R}^2)$ be L times weakly differentiable with all derivatives in $L^2(\mathbb{R}^2)$. Let furthermore ψ be a bounded compactly supported shearlet with M vanishing moments such that there exists an $\alpha \in (1/2, 1)$ with $(1 - \alpha) M \geq \frac{7}{4}$ and $\left(\alpha - \frac{1}{2}\right) L \geq \frac{7}{4}$.*

   i. *Let $\vec{\alpha}(t_0) = p \in \partial R$ be a regular point.*

      (a) *If s does not correspond to the normal direction of ∂R at p, then $\mathcal{SH}_\psi \chi_R(a, s_0, p)$ decays as*

$$\mathcal{SH}_\psi \chi_R(a, s, p) = \mathcal{O}\left(a^{(1-\alpha)M} + a^{(\alpha - \frac{1}{2})L}\right), \quad \textit{for all } \alpha \in \left(\frac{1}{2}, 1\right).$$

(b) If $\delta > 0$, $\|p - p_i\| > \delta$ for all corner points p_i, and $\tilde{s}$ corresponds to the normal direction of ∂R at p, then there exists a constant C_δ such that for all $a \in (0, 1]$

$$\lim_{a \to 0} a^{\frac{3}{4}} \int_{\widetilde{S}} \psi(x) \, \mathrm{d}x - C_\delta a^{\frac{5}{4}} \leq \mathcal{SH}_\psi \chi_R(a, s, p) \leq \lim_{a \to 0} a^{\frac{3}{4}} \int_{\widetilde{S}} \psi(x) \, \mathrm{d}x + C_\delta a^{\frac{5}{4}},$$

for $s \in B_a(\tilde{s})$, where

$$\widetilde{S} = \left\{ (x_1, x_2) \in \operatorname{supp} \psi \colon x_1 \leq \frac{1}{2\rho(s)^2} \left(\vec{\alpha}_1''(t_0) - s\vec{\alpha}_2''(t_0) \right) x_2^2 \right\}.$$

ii. Let p be a corner point.

(a) If $p \in \partial R$ is a corner point of the first type and $\tilde{s}$ corresponds to a normal direction of ∂R at p, then if $s \in B_a(\tilde{s})$

$$\lim_{a \to 0+} a^{-\frac{3}{4}} \mathcal{SH}_\psi \chi_R(a, s, p) \in \left\{ \int_{\widetilde{S}^{up}} \psi(x) \, \mathrm{d}x, \int_{\widetilde{S}^{down}} \psi(x) \, \mathrm{d}x \right\},$$

where

$$\widetilde{S}^{up} = \widetilde{S} \cap \{x \colon x_2 > 0\}, \quad \widetilde{S}^{down} = \widetilde{S} \cap \{x \colon x_2 < 0\}.$$

(b) If $p \in \partial R$ is a corner point of the first type and s does not correspond to a normal direction of ∂R at p, then

$$\mathcal{SH}_\psi \chi_R(a, s, p) = \mathcal{O}\left(a^{\frac{5}{4}} \right) \text{ for } a \to 0.$$

If furthermore $\psi(x_1, x_2) = \psi_1(x_1) \psi_2(x_2)$ for a wavelet $\psi_1 \in L^2(\mathbb{R})$, $\psi \in C^2(\mathbb{R}^2) \cap L^2(\mathbb{R}^2)$ and $\psi_2(0) = 0$, $\psi_2'(0) \neq 0$ and

$$\int_{(-\infty, 0)} \psi_1(x_1) x_1^2 \neq 0,$$

then

$$\lim_{a \to 0+} a^{-\frac{5}{4}} \left| \mathcal{SH}_\psi \chi_R(a, s, p) \right| > 0.$$

(c) Let $\psi(x_1, x_2) = \psi_1(x_1) \psi_2(x_2)$ with a bounded compactly supported wavelet ψ_1 and a compactly supported function $\psi_2 \in C^2(\mathbb{R}) \cap L^2(\mathbb{R})$ satisfying $\psi_2'(0) \neq 0$. If $p \in \partial R$ is a corner point of the second type and s does not correspond to a normal direction of ∂R at p, then

$$\mathcal{SH}_\psi \chi_R(a, s, p) = \mathcal{O}\left(a^{\frac{7}{4}} \right) \text{ for } a \to 0.$$

If furthermore ψ_1 has three vanishing moments and

$$\int_{(-\infty, 0)} \psi_1(x_1) x_1^3 \mathrm{d}x_1,$$

then

$$\lim_{a \to 0+} a^{-\frac{7}{4}} \left| \mathcal{SH}_\psi \chi_R(a, s, p) \right| > 0.$$

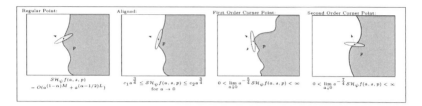

Figure 2.3: Illustration of the decay rates using compactly supported shearlets [78].

2.5   Digital Shearlets

Currently, there are several digital realizations of shearlets and their transform [35, 58, 61, 77, 79, 80, 86]. For all these approaches, except [58], the algorithms are publicly available for download. The implementations can be categorized into two groups. The underlying shearlet system is either based on band-limited functions or it is set up using functions with compact support in time domain. These two general categories will be further reviewed in the following subsections.

2.5.1   Band-limited Shearlets

As can be seen in the previous sections, the theoretical analysis of the shearlet transform is concentrated on the Fourier domain. Especially the cone-adapted discrete shearlet system is defined by a tiling of the frequency plane. Therefore, it seems natural to implement shearlets directly according to this definition. Consequently, the shearlets are computed in Fourier domain.

As described in [75], Easley et al. [35] provided the first numerical implementation of the discrete shearlet transform. On a given scale, the image is decomposed into a low pass and a high pass image by the Laplacian pyramid scheme. Next, a directional filtering is performed on the pseudo-polar grid. Finally, the Cartesian sampled values are reassembled and the inverse 2D FFT is applied. To reduce the Gibbs type ringing, i.e. overshoots near jump points of a signal, also a local variant of the shearlet transform has been implemented. For more details on the Gibbs phenomenon, see [119, Section 4.7]. The shearlet transform is then carried out by a convolution using directional filters governed by an approximation of the inverse shearlet transform of band-limited window functions. These filters were able to be implemented with a matrix representation that is of smaller size than the given image. Still they are not compactly supported in the traditional sense. The implementation is available at www.math.uh.edu/~dlabate.

Another approach is described in [79, 80]. Here, an isometric pseudo-polar Fourier transform is obtained by careful weighting of the pseudo-polar grid, for which its adjoint can be applied for the inverse transform. Band-limited shearlets are used to obtain tight frames such that the adjoint frame operator allows for reconstruction. The corresponding implementation is called *ShearLab* and available online in the form of a MATLAB toolbox at www.ShearLab.org.

In contrast to other approaches, the implementation of Häuser [60, 61], called *Fast Finite Shearlet Transform (FFST)*, utilizes a fully discrete setting. Not only the shearlet parameters a, s and t are discretized but also only a finite number of discrete translations is considered. The use of band-limited shearlets based on Meyer wavelets [93] enables the formation of a discrete Parseval frame. Therefore, a direct reconstruction of an image using formula (2.3) is possible. The FFST is also implemented as a MATLAB toolbox and online available at www.mathematik.uni-kl.de/imagepro/software/ffst.

On the one hand, a key advantage of a digitalization of the discrete shearlet transform in the Fourier domain is the possibility to easily construct tight or even Parseval frames by using band-limited shearlets. On the other hand, it entails the drawback that the band-limited shearlets have infinite support in time domain. This introduces the above mentioned Gibbs type ringing. Considering the shearlet transform as a convolution in time domain, one can intuitively imagine that the support size has a major effect on the quality of localizing structures such as edges in images. In case of an infinite shearlet support, image points outside of a near neighborhood are taken into account during the computation of the shearlet transform of a given image point. Figure 2.4a shows a shearlet in time domain for a 128×64 image on a fine scale generated with the FFST toolbox. It illustrates that, even under a fine scale, still image points in a wider neighborhood have a significant impact on the value of the shearlet coefficients using this shearlet.

2.5.2 Compactly Supported Shearlets

In contrast to the approaches above, some implementations use compactly supported instead of band-limited shearlets. The main reason is to provide better localization of the shearlet transform which is useful for several applications.

The first implementation doing so was provided by Lim [86]. The presented shearlets are generated by separable functions, while each function component is compactly supported. These separable functions are constructed using a Multi Resolution Analysis, which then leads to the discrete shearlet transform to compute the shearlet coefficients of an image. As a main drawback, the corresponding shearlet system does not provide a tight frame as band-limited shearlets are able to.

In [87], Lim was able to improve this approach in regards to frame properties by using non-separable compactly supported shearlet mother functions. It was found that non-separable compactly supported functions can better approximate band-limited mother functions, which are able to lead to Parseval frames. The implementation of this approach is available online as part of the *ShearLab 3D* MATLAB toolbox [77]. As its name implies this toolbox extends [87] to the 3D situation. Furthermore, it utilizes *universal* shearlets which allow more flexibility in terms of scaling. That means that at each scaling level the utilization of a different type of scaling is possible by the introduction of a scale dependent scaling matrix

$$A^j_{\alpha_j} := \begin{pmatrix} a & 0 \\ 0 & a^{\alpha_j} \end{pmatrix}.$$

The toolbox is downloadable at www.ShearLab.org.

In Figure 2.4b, a shearlet at a fine scale generated by the ShearLab3D toolbox for a 128×64 image is shown. It has to be mentioned that it was not possible to generate a shearlet system with the default parameters for such a small sized image. Instead the parameters had to be adapted to get shearlets with support as small as possible with the side effect of losing accuracy. One may argue that the image size of 128×64 is unreasonable small. However, algorithms for pedestrian detection, which are our main concern in this thesis, are trained on sample images of size 128×64 or 64×32. Certainly, it is possible to pad the images in the training process such that other parameters may be used. Anyway, this would lead to a bigger shearlet support which is not beneficial for detecting such small object instances as required. For example in the Caltech benchmark, the median height of pedestrians is just 48 pixels.

As can be seen in Figure 2.4, the amount of shearlet pixels with significant absolute value involved in the computation of the shearlet coefficients at a given image point is clearly smaller

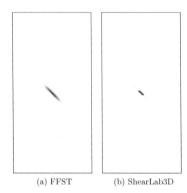

(a) FFST   (b) ShearLab3D

Figure 2.4: Fine scale shearlet in time domain generated for a 128×64 image.

for ShearLab 3D shearlets as for the FFST. Intuitively, this leads to advantages for the task of feature extraction used for pedestrian detection. Still, the shearlet filter sizes generated by the ShearLab 3D toolbox seem to have room for improvement in regards to accuracy of localizing edges in images. In the next chapter, we will set up our own shearlets with the aim to generate shearlets with smallest possible support while retaining accuracy in directional adjustment.

2.6 Conclusion

In this chapter, we examined the current contributions on the shearlet framework. First, we provided theoretical key results which will serve as a base for our own work. Second, we analyzed current available realizations with a focus on their capabilities for image feature extraction.

We observed that most implementations concentrate them on fulfilling tight or even Parseval frame properties instead of giving the ability to compute high quality image features. In order to yield tight or Parseval frames, band-limited mother functions are commonly chosen. As a matter of fact, the resulting shearlets have infinite support in time domain. Therefore, the localization of edges in images using these shearlets is improvable.

However, we found that state-of-the-art realizations applying compactly supported shearlets, such as ShearLab 3D [77], still generate shearlet filters with considerably large pixel size, which has a negative effect on the localization of edges. In order to compute high quality image features by precisely capturing edges in images, we set up our own shearlet design which we will describe in the subsequent chapter.

"Inspiration is hard to come by. You have to take it where you find it."

Bob Dylan

3

Local Precision Shearlets

In this chapter, we set up a new shearlet design tailored to the need that we have to detect pedestrian instances with very small pixel size. In the previous chapter, we found that currently available shearlet implementations are not fitted to that need since their shearlet filters have a relatively big pixel size. Therefore, our shearlet design focuses on the ability to provide accurate shearlet filters with small pixel size.

First, we define a new type of compactly supported shearlets. We design the mother shearlet function specifically in order to be able to precisely localize structures in images. Second, we define a new shearlet system. Our shearlet system is significantly different to common shearlet systems in regards to directional distribution of the shearlets and the flexibility in defining the used shearlets per scale. Subsequently, we show that our shearlet system forms a frame for $L^2(\mathbb{R}^2)$. Finally, we discuss the possible algorithms for a signal reconstruction and the practical application of our shearlets. Parts of this chapter have been published in [107].

3.1 Mother Shearlet

The choice of frequency bounded functions makes it possible to design shearlets that form a Parseval frame for $L^2(\mathbb{R}^2)$ such that a reconstruction of the signal given the shearlet coefficients is possible. Besides its advantages, this approach leaves the drawback that the resulting functions $\psi_{a,s,t}$ have infinite support in time domain resulting in improvable edge detections. To avoid this disadvantage for the localization of structures in images, which is crucial for feature extraction, we design the mother shearlet ψ such that it has compact support $[-b_1, b_1] \times [-b_2, b_2] \subset \mathbb{R}^2$ in time domain with support boundaries $b_1, b_2 \in \mathbb{R}^+$ and $b_2 > b_1$. The last condition guarantees that we have elongated functions already at the coarsest scale of the shearlet system. In addition, we follow the approach of [69, 80, 86] and set up our mother shearlet as a separable function. As preparation, we will define the basic function components of our mother shearlet.

Definition 3.1 ([90]). We call a function $\psi_1 \colon \mathbb{R} \to \mathbb{C}$ *mother wavelet* if it is of zero average, i.e.

$$\int_{-\infty}^{\infty} \psi_1(x)\, dx = 0.$$

Definition 3.2 ([83]). Let $U \subset \mathbb{R}$ be a closed subset and $B \subset \mathbb{R}$ be a set with $U \subset B$. A continuous function $\psi_2 \colon \mathbb{R} \to \mathbb{R}$ is called a *bump function* for U supported in B if $0 \leq \psi_2(x) \leq 1$ for $x \in \mathbb{R}$, $\psi_2(x) \equiv 1$ for $x \in U$ and supp $\psi_2 \subseteq B$.

More precisely, we set up our mother shearlet $\psi \in L^2(\mathbb{R}^2)$ according to the following definition.

Definition 3.3. Let $B_1 = [-b_1, b_1]$ and $B_2 = [-b_2, b_2]$ with $b_1, b_2 \in \mathbb{R}^+$, $b_2 > b_1$ and $U \subset B_2$. For $x \in \mathbb{R}^2$ the mother shearlet given by

$$\psi(x_1, x_2) := \psi_1(x_1)\,\psi_2(x_2) \tag{3.1}$$

is called a *local precision shearlet* if it fulfills the following conditions

   i. $\psi_1 \in L^1(\mathbb{R}) \cap L^2(\mathbb{R})$ is a mother wavelet compactly supported in B_1 and point-symmetric to $(0, 0)$.

   ii. $\lambda\psi_2 \in L^1(\mathbb{R}) \cap L^2(\mathbb{R})$, $\lambda > 0$, is a bump function for U supported in B_2 axis-symmetric to $x_2 = 0$.

The separability (3.1) allows easy evaluations of theoretical properties as well as it enables simple algorithmic realization. Concerning the Fourier transform $\hat{\psi} := \mathcal{F}(\psi)$ of a local precision shearlet ψ we have

$$
\begin{aligned}
\hat{\psi}(\xi_1, \xi_2) &= \int_{\mathbb{R}^2} \psi_1(x_1)\,e^{-2\pi i x_1 \xi_1} \psi_2(x_2)\,e^{-2\pi i x_2 \xi_2}\mathrm{d}x_1\mathrm{d}x_2 \\
&= \int_{\mathbb{R}} \psi_1(x_1)\,e^{-2\pi i x_1 \xi_1}\mathrm{d}x_1 \int_{\mathbb{R}} \psi_2(x_2)\,e^{-2\pi i x_2 \xi_2}\mathrm{d}x_2 \\
&= \hat{\psi}_1(\xi_1)\,\hat{\psi}_2(\xi_2). \tag{3.2}
\end{aligned}
$$

The definition of our ψ_1-component is motivated by the work of Mallat and Zhong [91], where a wavelet is chosen to be the first derivative of a *smoothing function* θ. Here, a smoothing function is defined as any function $\theta \colon \mathbb{R} \to \mathbb{R}$ that has an integral equal to 1 and that converges to 0 at infinity. As an example, one can use a Gaussian as smoothing function. In that way this type of wavelet is sharing the property of ψ_1 being a point-symmetric function. The corresponding wavelet transform of a signal $f \in L^2(\mathbb{R})$ is then the first derivative of the signal smoothed at the corresponding scale of the wavelet. As Mallat and Zhong [91] point out, the detection of local extrema of the corresponding wavelet coefficients conforms to the one-dimensional Canny edge detection [16] if one chooses θ to be a Gaussian. For a fast algorithmic implementation, they introduce a wavelet which is set up as first derivative of a cubic B-spline. Duval-Poo et al. [34] were able to improve edge detection with the FFST shearlet implementation [60, 61] by using this wavelet as a replacement for the originally used Meyer wavelet [93] in the ψ_1-component. Our definition of the ψ_2-component guarantees the symmetry in $\mathbb{R}^2$. Li and Shen [85] used symmetric shearlets based on B-splines which provide optimally sparse approximations of cartoon-like images.

We use these ideas as the basis to define a shearlet based on B-splines, which is fulfilling the conditions of a local precision shearlet. In contrast to [34], our approach results in a separable function which uses B-splines not just in ψ_1 but also in ψ_2. Since Duval-Poo et al. [34] use the ψ_2-component of the FFST [60, 61] along with non-separable mother functions, the resulting shearlets are not compactly supported in time domain. Furthermore, we extend the approach to allow higher orders of derivatives. The resulting shearlet is called *spline shearlet* and is set up as a centralized B-spline in the bump component ψ_2 and its q-th derivative in the wavelet component ψ_1. The cardinal B-spline $N_p(x) \colon \mathbb{R} \to \mathbb{R}$ of order $p \geq 2$ is defined by

$$N_p(x) := (N_{p-1} * N_1)(x) \tag{3.3}$$

with

$$N_1(x) := \begin{cases} 1 & 0 \leq x < 1, \\ 0 & \text{otherwise.} \end{cases}$$

Before stating the definition of a spline shearlet, we first present some useful properties of cardinal B-splines.

Theorem 3.4 ([18]). *The cardinal B-spline N_p of order $p \in \mathbb{N}$ satisfies the following properties:*

   i. $\operatorname{supp} N_p = [0, p]$.

   ii. $N_p(x) > 0$ for all $0 < x < p$.

   iii. $\sum_{k=-\infty}^{\infty} N_p(x - k) = 1$ for all $x \in \mathbb{R}$.

   iv. $N_p'(x) = N_{p-1}(x) - N_{p-1}(x - 1)$ for all $x \in \mathbb{R}$.

   v. The cardinal B-splines N_p and N_{p-1} are related by the identity

$$N_p(x) = \frac{x}{p-1} N_{p-1}(x) + \frac{p-x}{p-1} N_{p-1}(x-1).$$

   vi. N_p is symmetric with respect to the center of its support, i.e.

$$N_p\left(\frac{p}{2} + x\right) = N_p\left(\frac{p}{2} - x\right), \quad x \in \mathbb{R}.$$

Definition 3.5. Let $N_p \in L^2(\mathbb{R})$ be a cardinal B-spline of order $p \in \mathbb{N}$. For $q \in \mathcal{Q}_m$ with

$$\mathcal{Q}_m := \{1, 3, \ldots, p - (1 + \operatorname{mod}(p, 2))\}$$

and $r > 1$, we call $\psi(x_1, x_2) = \psi_1(x_1)\psi_2(x_2)$ a *spline shearlet of order p and q-th derivative* if

$$\psi_1(x) = \widetilde{N}_p^{(q)}(rx) \tag{3.4}$$

and

$$\psi_2(x) = \widetilde{N}_p(x) \tag{3.5}$$

with $x \in \mathbb{R}$ and $\widetilde{N}_p(x) := N_p(x + p/2)$.

It is easy to observe that the spline shearlet fulfills the conditions of a local precision shearlet. The constant $r > 1$ causes that we have $b_2 > b_1$, i.e. that the support of ψ_2 is larger than the one of ψ_1. Since $\widetilde{N}_p(x)$ and $\widetilde{N}_p^{(q)}(x)$ have the same support, r equals the ratio b_2/b_1. In other words, for the support boundaries of the resulting shearlet we get $b_1 = p/2r$ and $b_2 = p/2$. Figure 3.1 shows a spline shearlet with $p = 5$, $q = 1$ and $r = 3/2$.

Next, we examine the number of vanishing moments of the first component of a spline shearlet. As we will see in Chapter 4, the characterization of edge points using a local precision shearlet is depending on the number of vanishing moments of ψ_1.

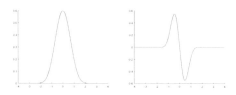

(a) Centralized B-spline ψ_2 with $p = 5$ as well as its derivative ψ_1 compressed with $r = {}^3/_2$.

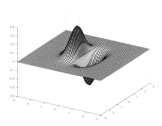

(b) Spline shearlet $\psi(x) = \psi_1(x_1)\psi_2(x_2)$.

Figure 3.1: Example of the spline mother shearlet and its components for $p = 5$.

Lemma 3.6. *Let $\psi(x_1, x_2) = \psi_1(x_1)\psi_2(x_2)$ be a spline shearlet of order p and q-th derivative. Then ψ_1 has q vanishing moments, i.e.*

$$\int_{\mathbb{R}} \psi_1(x_1)\, x_1^l \mathrm{d}x_1 = 0$$

for all $l < q$.

Proof. For a spline shearlet of order p and q-th derivative we have

$$\psi_1(x_1) = \tilde{N}_p^{(q)}(rx_1)$$

with $q \in 2\mathbb{N} - 1$, $r > 1$ and $\tilde{N}_p^{(q)}(x_1) = N_p^{(q)}(x_1 + {}^p/_2)$, where N_p is a cardinal B-spline of order $p \in \mathbb{N}$. Furthermore we have $b_1 = {}^p/_{2r}$. Integration by parts provides

$$
\begin{aligned}
\int_{-p/2r}^{p/2r} \tilde{N}_p^{(q)}(rx_1)\, x_1^l \mathrm{d}x_1 &= \frac{1}{r}\bigg(\underbrace{\tilde{N}_p^{(q-1)}({}^p/_2)}_{=0} \cdot ({}^p/_{2r})^l - \underbrace{\tilde{N}_p^{(q-1)}(-{}^p/_2)}_{=0} \cdot (-{}^p/_{2r})^l \\
&\qquad - \int_{-p/2r}^{p/2r} \tilde{N}_p^{(q-1)}(rx_1)\, lx^{l-1} \mathrm{d}x_1 \bigg) \\
&= -\frac{l}{r} \int_{-p/2r}^{p/2r} \tilde{N}_p^{(q-1)}(rx_1)\, x_1^{l-1} \mathrm{d}x_1.
\end{aligned}
$$

Repeating this procedure l times with $l < q$, we have

$$\int_{-p/2r}^{p/2r} \tilde{N}_p^{(q)}(rx_1)\, x_1^l \mathrm{d}x_1 = \left(-\frac{l}{r}\right)^l \int_{-p/2r}^{p/2r} \tilde{N}_p^{(q-l)}(rx_1)\, \mathrm{d}x_1$$

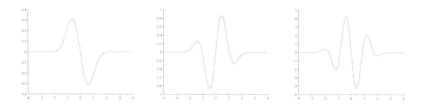

Figure 3.2: Shearlet component ψ_1 of a spline shearlet of order $p = 9$ and $q = 1$, $q = 3$ and $q = 5$.

$$= \left(-\frac{l}{r}\right)^l \left(\underbrace{\tilde{N}_p^{(q-l-1)}\left(p/2\right)}_{=0} - \underbrace{\tilde{N}_p^{(q-l-1)}\left(p/2\right)}_{=0}\right)$$

$$= 0.$$

For $l = q$, we have

$$\left(-\frac{l}{r}\right)^q \int_{-p/2r}^{p/2r} \tilde{N}_p\left(rx_1\right) x_1^{l-q} \mathrm{d}x_1 = \left(-\frac{l}{r}\right)^q \int_{-p/2r}^{p/2r} \tilde{N}_p\left(rx_1\right) \mathrm{d}x_1$$

$$\neq 0,$$

due to the properties of a B-spline described in Theorem 3.4. □

A big advantage of spline shearlets is that one can choose the grade of smoothness by the order of the B-spline as well as the number of vanishing moments by the choice which derivative to take. Figure 3.2 shows ψ_1 of a spline shearlet of order $p = 9$ with different numbers of vanishing moments. An increase in vanishing moments automatically increases oscillations. As we will present in Section 4.2, in an edge detection algorithm these oscillations create artifacts reducing the detection precision.

Another advantage is that a spline shearlet has explicit closed form expressions in time as well as in frequency domain. For the Fourier transform of its centralized version $\tilde{N}_1$ one gets

$$\mathcal{F}\left(\tilde{N}_1\right)(\xi) = \int_{-1/2}^{1/2} e^{-2\pi i x\xi} \mathrm{d}x$$

$$= -\left[\frac{e^{-2\pi i x\xi}}{2\pi i \xi}\right]_{-1/2}^{1/2}$$

$$= \frac{1}{\pi\xi}\frac{1}{2i}\left(e^{\pi i\xi} - e^{-\pi i\xi}\right)$$

$$= \frac{1}{\pi\xi}\sin\left(\pi\xi\right)$$

$$= \operatorname{sinc}\left(\xi\right).$$

For a description on basic techniques concerning Fourier analysis applied here, see Appendix B. For $\hat{\psi}_2\left(\xi\right) := \mathcal{F}(\psi_2)\left(\xi\right) = \mathcal{F}(\tilde{N}_p)\left(\xi\right)$ we obtain using (3.3)

$$\hat{\psi}_2\left(\xi\right) = \operatorname{sinc}^{p-1}\left(\xi\right). \tag{3.6}$$

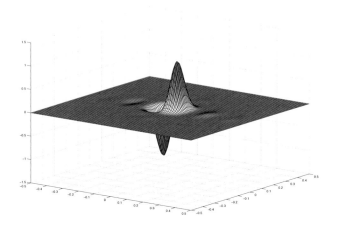

Figure 3.3: The spline shearlet in frequency domain

Since ψ_1 is a derivative of centralized B-spline, we get

$$\hat{\psi}_1\left(\xi\right) = \frac{1}{r}\left(i2\pi\xi\right)^q \operatorname{sinc}^{p-1}\left(\frac{\xi}{r}\right). \tag{3.7}$$

Corollary 3.7. *The Fourier transform of a spline shearlet ψ of order $p = n + 1$, and q-th derivative, defined by (3.4) and (3.5) has a closed-form expression given by*

$$\hat{\psi}\left(\xi_1, \xi_2\right) = \frac{1}{r}\left(i2\pi\frac{\xi_1}{r}\right)^q \operatorname{sinc}^n\left(\frac{\xi_1}{r}\right)\operatorname{sinc}^n\left(\xi_2\right).$$

Proof. According to (3.2) we have $\hat{\psi}\left(\xi_1, \xi_2\right) = \hat{\psi}_1\left(\xi_1\right)\hat{\psi}_2\left(\xi_2\right)$. Using (3.6) and (3.7) shows the result.   □

In time domain, one can express a cardinal B-spline as a composition of piecewise polynomials. That means, a cardinal B-spline of order p is at each interval $[k, k+1]$, $0 \leq k \leq p-1$, a polynomial of degree $p-1$. The coefficients of the involved polynomials can be calculated by a simple algorithm [95]. For example, for $p = 3$ we have the Quadratic B-spline

$$N_3\left(x\right) = \begin{cases} \frac{1}{2}x^2 & \text{for } 0 \leq x < 1, \\ -x^2 + 3x - \frac{3}{2} & \text{for } 1 \leq x < 2, \\ \frac{1}{2}x^2 - 3x + \frac{9}{2} & \text{for } 2 \leq x < 3, \\ 0 & \text{otherwise.} \end{cases}$$

Given the cardinal B-spline by piecewise polynomials, the calculation of its q-th derivative is very simple.

There are several other ways to define shearlets based on splines. One possibility is to define ψ_1 by a compactly supported B-spline wavelet [19, 20], which is given by the following definition.

Definition 3.8 ([20]). Let $N_p \in L^2(\mathbb{R})$ be a cardinal B-spline of order $p \in \mathbb{N}$. Then the compactly supported B-spline wavelet $\psi_{C,p}$ of order p is defined by

$$\psi_{C,p}(x) := \frac{1}{2^{2p-1}} \sum_{j=0}^{2p-2} (-1)^j N_{2p}(j+1) N_{2p}^{(p)}(2x-j). \tag{3.8}$$

For $p = 1$ we have the well-known Haar function

$$\psi_{C,1} := \begin{cases} 1 & \text{for } 0 \le x < \frac{1}{2}, \\ -1 & \text{for } \frac{1}{2} \le x < 1, \\ 0 & \text{otherwise.} \end{cases}$$

According to Chui and Wang [20], $\psi_{C,p}$ is axis-symmetric for even p and point-symmetric for odd p. Furthermore they state that supp $\psi_{C,p} = [0, 2p-1]$ and thus for a centralized version $\tilde{\psi}_{C,p}(x) = \psi_{C,p}(x + {(2p-1)}/{2})$ we have supp $\tilde{\psi}_{C,p} = [-{(2p-1)}/{2}, {(2p-1)}/{2}]$. Figure 3.4a shows B-spline wavelets of orders $p = 1$, $p = 3$ and $p = 5$. As can be seen, an increase in the order of the B-spline directly increases the oscillations of the wavelet.

Another way to set up shearlets based on splines is presented by Lim [86]. Here, a pair of shearlets is defined in frequency domain by

$$\hat{\psi}_0^1 := (i)^l (\sin(\pi\xi_1))^l \, \hat{\theta}_p(\xi_1) \, \hat{\theta}_p(\xi_2)$$

$$\hat{\psi}_0^2 := (i)^l \left(\sin\left(\frac{\pi\xi_1}{2}\right)\right)^l \, \hat{\theta}_p\left(\frac{\xi_1}{2}\right) \hat{\theta}_p(\xi_2),$$

where $\hat{\theta}_p$ is the Fourier transform a box spline of order p given by

$$\hat{\theta}(\xi_1) = \left(\frac{\sin(\pi\xi_1)}{\pi\xi_1}\right)^{p+1} e^{-i\epsilon\pi\xi_1}, \text{ with } \epsilon = \begin{cases} 1 & \text{if } p \text{ is even,} \\ 0 & \text{if } p \text{ is odd.} \end{cases}$$

The x_1 component of this shearlet pair shows a very similar shape as the one of a spline shearlet as illustrated in Figure 3.4b. Also here, we can adjust smoothness and oscillations separately. But in contrast to spline shearlets, we do not have a closed form expression in time domain. Concerning practical application, one either needs to compute the shearlet transform in Fourier domain and the result has to be transferred to time domain by the inverse Fourier transform. This leads to higher computational complexity in case of small filters, see 3.5. Or the shearlets have to be transferred to the time domain such that the shearlet transform can be performed by a convolution. Both ways entail a potential loss of calculation accuracy due to the involvement of the inverse Fourier transform.

Turning to the properties of a general local precision shearlet, since ψ_1 is an odd function and ψ_2 is an even one, [119] provides the following properties of the Fourier transform of ψ.

Lemma 3.9. Let $\psi \in L^2(\mathbb{R}^2)$ be a local precision shearlet. Then $\hat{\psi}_1$ is an odd function, i.e.

$$\hat{\psi}_1(\xi) = -\hat{\psi}_1(-\xi) \tag{3.9}$$

and $\hat{\psi}_2$ is even for all $\xi \in \mathbb{R}$. Furthermore, $\hat{\psi}_1$ is purely imaginary and $\hat{\psi}_2$ is real.

With these properties, we get the following result for the continuous shearlet transform $\mathcal{SH}_\psi$ described in Chapter 2.

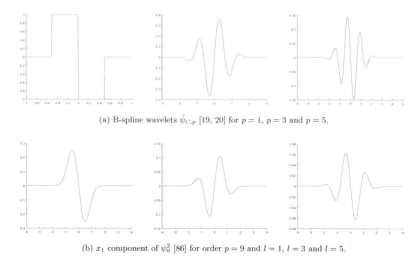

(a) B-spline wavelets $\tilde{\psi}_{C,p}$ [19, 20] for $p = 1$, $p = 3$ and $p = 5$.

(b) x_1 component of ψ_0^2 [86] for order $p = 9$ and $l = 1$, $l = 3$ and $l = 5$.

Figure 3.4: Shearlet components in x_1 direction for different spline approaches.

Theorem 3.10. *Let* $\psi \in L^2\left(\mathbb{R}^2\right)$ *be a local precision shearlet that fulfills the admissibility condition (2.4). Then for*

$$c_\psi^- = \int_\mathbb{R} \int_{-\infty}^0 \frac{\left|\hat{\psi}\left(\xi_1, \xi_2\right)\right|^2}{\xi_1^2} \mathrm{d}\xi_1 \mathrm{d}\xi_2 \quad and \quad c_\psi^+ = \int_\mathbb{R} \int_0^\infty \frac{\left|\hat{\psi}\left(\xi_1, \xi_2\right)\right|^2}{\xi_1^2} \mathrm{d}\xi_1 \mathrm{d}\xi_2$$

we have $c_\psi^- = c_\psi^+ = c_\psi$ *and* $\mathcal{SH}_\psi$ *is a* c_ψ*-multiple of an isometry.*

Proof. According to (3.2) and (3.9) we get

$$
\begin{aligned}
c_\psi^- &= \int_\mathbb{R} \int_{-\infty}^0 \frac{\left|\hat{\psi}\left(\xi_1, \xi_2\right)\right|^2}{\xi_1^2} \mathrm{d}\xi_1 \mathrm{d}\xi_2 \\
&= \int_\mathbb{R} \int_{-\infty}^0 \frac{\left|\hat{\psi}_1\left(\xi_1\right) \hat{\psi}_2\left(\xi_2\right)\right|^2}{\xi_1^2} \mathrm{d}\xi_1 \mathrm{d}\xi_2 \\
&= \int_\mathbb{R} \int_{-\infty}^0 \frac{\left|-\hat{\psi}_1\left(-\xi_1\right) \hat{\psi}_2\left(\xi_2\right)\right|^2}{\xi_1^2} \mathrm{d}\xi_1 \mathrm{d}\xi_2 \\
&= \int_\mathbb{R} \int_0^\infty \frac{\left|\hat{\psi}_1\left(\xi_1\right) \hat{\psi}_2\left(\xi_2\right)\right|^2}{\xi_1^2} \mathrm{d}\xi_1 \mathrm{d}\xi_2.
\end{aligned}
$$

Therefore we have $c_\psi^- = c_\psi^+ = c_\psi$. The isometry statement immediately follows from Theorem 2.7. $\qquad\square$

Besides the closed-form expression of a spline shearlet in frequency domain, Corollary 3.7 provides the following result.

Theorem 3.11. *A spline shearlet ψ of order $p = n + 1$ with q-th derivative is admissible and we have $c_\psi^+ = c_\psi^- = c_\psi$ with*

$$c_\psi^+ = \frac{4^q \pi^2}{r^5 2^{2q} (2n - 2q + 1)! (2n - 1)!} \left\langle \begin{array}{c} 2n - 1 \\ n - 1 \end{array} \right\rangle \sum_{j=0}^{n} (-1)^j \left(\begin{array}{c} 2n \\ j \end{array} \right) (n - j)^{2n - 2q + 1}.$$

and the Eulerian number

$$\left\langle \begin{array}{c} k \\ l \end{array} \right\rangle = \sum_{j=0}^{l+1} (-1)^j \left(\begin{array}{c} k + 1 \\ j \end{array} \right) (l - j + 1)^k \quad \text{for } k, l \in \mathbb{N}.$$

Proof. According to Corollary 3.7 we obtain

$$
\begin{aligned}
c_\psi^+ &= \int_\mathbb{R} \int_0^\infty \frac{\left| \frac{1}{r} \left(\frac{i 2\pi}{r} \xi_1 \right)^q \operatorname{sinc}^n \left(\frac{\xi_1}{r} \right) \operatorname{sinc}^n (\xi_2) \right|^2}{\xi_1^2} \mathrm{d}\xi_1 \mathrm{d}\xi_2 \\
&= \int_\mathbb{R} \int_0^\infty \frac{\frac{1}{r^2} \left(\frac{2\pi}{r} \xi_1 \right)^{2q} \operatorname{sinc}^{2n} \left(\frac{\xi_1}{r} \right) \operatorname{sinc}^{2n} (\xi_2)}{\xi_1^2} \mathrm{d}\xi_1 \mathrm{d}\xi_2 \\
&= 4^q \int_\mathbb{R} \int_0^\infty \frac{\pi^{2q}}{r^{2q+2}} \xi_1^{2q-2} \operatorname{sinc}^{2n} \left(\frac{\xi_1}{r} \right) \operatorname{sinc}^{2n} (\xi_2) \, \mathrm{d}\xi_1 \mathrm{d}\xi_2 \\
&= 4^q \int_\mathbb{R} \int_0^\infty \frac{\pi^{2q}}{r^{2q+2}} \xi_1^{2q-2} \left(\frac{\sin \left(\frac{\pi}{r} \xi_1 \right)}{\frac{\pi}{r} \xi_1} \right)^{2n} \left(\frac{\sin (\pi \xi_2)}{\pi \xi_2} \right)^{2n} \mathrm{d}\xi_1 \mathrm{d}\xi_2 \\
&= 4^q \int_0^\infty \frac{\sin^{2n} \left(\frac{\pi}{r} \xi_1 \right)}{\frac{\pi^{2(n-q)}}{r^{2(n-q-1)}} \xi_1^{2(n-q+1)}} \mathrm{d}\xi_1 \int_\mathbb{R} \left(\frac{\sin (\pi \xi_2)}{\pi \xi_2} \right)^{2n} \mathrm{d}\xi_2
\end{aligned}
$$

and therefore

$$c_\psi^+ = \frac{4^q \pi^2}{r^4} \int_0^\infty \frac{\sin^{2n} \left(\frac{\pi}{r} \xi_1 \right)}{\left(\frac{\pi}{r} \xi_1 \right)^{2(n-q+1)}} \mathrm{d}\xi_1 \int_\mathbb{R} \left(\frac{\sin (\pi \xi_2)}{\pi \xi_2} \right)^{2n} \mathrm{d}\xi_2. \qquad (3.10)$$

From [120, page 2703] we know

$$
\begin{aligned}
\int_0^\infty \frac{\sin^{2n} (\xi_1)}{(\xi_1)^{2(n-q+1)}} &= \frac{\pi}{2^{2n} (2 (n - q + 1) - 1)!} \sum_{j=0}^{n} (-1)^j \left(\begin{array}{c} 2n \\ j \end{array} \right) (2n - 2j)^{2(n-q+1)-1} \\
&= \frac{\pi 2^{2n-2q+1}}{2^{2n} (2n - 2q + 1)!} \sum_{j=0}^{n} (-1)^j \left(\begin{array}{c} 2n \\ j \end{array} \right) (n - j)^{2n - 2q + 1} \\
&= \frac{\pi}{2^{2q-1} (2n - 2q + 1)!} \sum_{j=0}^{n} (-1)^j \left(\begin{array}{c} 2n \\ j \end{array} \right) (n - j)^{2n - 2q + 1}
\end{aligned}
$$

and

$$\int_0^\infty \left(\frac{\sin (\xi_1)}{\xi_1} \right)^{2n} \mathrm{d}\xi_1 = \frac{\pi}{2 (2n - 1)!} \left\langle \begin{array}{c} 2n - 1 \\ n - 1 \end{array} \right\rangle.$$

Integration by substitution yields

$$\int_0^\infty \frac{\sin^{2n} \left(\frac{\pi}{r} \xi_1 \right)}{\left(\frac{\pi}{r} \xi_1 \right)^{2(n-q+1)}} \mathrm{d}\xi_1 = \frac{1}{r 2^{2q-1} (2n - 2q + 1)!} \sum_{j=0}^{n} (-1)^j \left(\begin{array}{c} 2n \\ j \end{array} \right) (n - j)^{2n - 2q + 1}$$

and

$$\int_0^\infty \left(\frac{\sin (\pi \xi_1)}{\pi \xi_1} \right)^{2n} \mathrm{d}\xi_1 = \frac{1}{2 (2n - 1)!} \left\langle \begin{array}{c} 2n - 1 \\ n - 1 \end{array} \right\rangle.$$

It follows that

$$
\begin{aligned}
c_\psi^+ &= \frac{4^q \pi^2}{r^5 2^{2q-1} (2n - 2q + 1)!} \left(\sum_{j=0}^{n} (-1)^j \binom{2n}{j} (n-j)^{2n-2q+1} \right) \frac{1}{2(2n-1)!} \left\langle \begin{array}{c} 2n-1 \\ n-1 \end{array} \right\rangle \\
&= \frac{4^q \pi^2}{r^5 2^{2q} (2n - 2q + 1)! \, (2n-1)!} \left\langle \begin{array}{c} 2n-1 \\ n-1 \end{array} \right\rangle \sum_{j=0}^{n} (-1)^j \binom{2n}{j} (n-j)^{2n-2q+1}.
\end{aligned}
$$

In particular, we get $c_\psi^+ < \infty$. According to Theorem 3.10 we have $c_\psi^+ = c_\psi^- = c_\psi$ showing the result. $\qquad\square$

In case of a spline shearlet with $q = 1$, the formula for c_ψ simplifies to the following result.

Corollary 3.12. *A spline shearlet ψ of order $p = n + 1$ with 1-st derivative is admissible and we have*

$$
c_\psi^+ = c_\psi^- = c_\psi = \frac{2}{r} \left(\frac{\pi}{r^2 (2n-1)!} \left\langle \begin{array}{c} 2n-1 \\ n-1 \end{array} \right\rangle \right)^2.
$$

Proof. For $q = 1$ we get for (3.10)

$$
c_\psi^+ = \frac{4\pi^2}{r^4} \int_0^\infty \left(\frac{\sin(\pi\xi_1)}{\pi\xi_1} \right)^{2n} d\xi_1 \int_{\mathbb{R}} \left(\frac{\sin(\pi\xi_2)}{\pi\xi_2} \right)^{2n} d\xi_2
$$

and therefore

$$
\begin{aligned}
c_\psi^+ &= \frac{4\pi^2}{r^4} \frac{1}{2r(2n-1)!} \left\langle \begin{array}{c} 2n-1 \\ n-1 \end{array} \right\rangle \frac{1}{(2n-1)!} \left\langle \begin{array}{c} 2n-1 \\ n-1 \end{array} \right\rangle \\
&= \frac{2}{r} \left(\frac{\pi}{r^2 (2n-1)!} \left\langle \begin{array}{c} 2n-1 \\ n-1 \end{array} \right\rangle \right)^2.
\end{aligned}
$$

$\qquad\square$

For the continuous shearlet transform, we derive the following finding.

Corollary 3.13. *Let $\psi \in L^2(\mathbb{R}^2)$ be a spline shearlet of order p and q-th derivative. Then $\mathcal{SH}_\psi$ is a c_ψ-multiple of an isometry with*

$$
c_\psi = \frac{4^q \pi^2}{r^5 2^{2q} (2n - 2q + 1)! \, (2n-1)!} \left\langle \begin{array}{c} 2n-1 \\ n-1 \end{array} \right\rangle \sum_{j=0}^{n} (-1)^j \binom{2n}{j} (n-j)^{2n-2q+1}.
$$

Proof. The result directly follows from Theorems 3.10 and 3.11. $\qquad\square$

Theorem 3.14. *Let Γ_R be given by (2.8) with $a > 1$ and $b = 1$, set $\beta > 0$ and $\gamma > 2(\beta + 2)$ and assume that $\beta' \geq \beta + \gamma$ and $\gamma' \geq \beta' - \beta + \gamma$. If $\psi \in L^2(\mathbb{R}^2)$ is a spline shearlet of order $p = n + 1$ and q-th derivative with $n \geq \gamma' - \beta' + p$ and $q \geq \beta'$, then there exists $c_0 > 0$ such that the regular shearlet system $SH(\psi, \Gamma_R)$ is a frame for $L^2(\mathbb{R}^2)$ for all $c \leq c_0$.*

Proof. We want to show that the conditions of Theorem 2.13 hold true for a spline shearlet with $n \geq \gamma' - \beta' + p$ and $q \geq \beta'$. Considering the condition (2.13), we set

$$
d_{\hat{\psi}_1}(\xi_1) := \left| \hat{\psi}_1(\xi_1) \right| \frac{\left(1 + |\xi_1|^2 \right)^{\gamma'/2}}{|\xi_1|^{\beta'}}.
$$

Using this notation we have

$$
\begin{aligned}
d_{\hat{\psi}_1}(\xi_1) &= \left| \frac{1}{r} \left(\frac{i2\pi}{r} \xi_1 \right)^q \operatorname{sinc}^n \left(\frac{\xi_1}{r} \right) \right| \frac{\left(1 + |\xi_1|^2 \right)^{\gamma'/2}}{|\xi_1|^{\beta'}} \\
&= \frac{(2\pi)^q}{r^{q+1}} \left| \xi_1^q \left(\frac{\sin \left(\frac{\pi}{r} \xi_1 \right)}{\frac{\pi}{r} \xi_1} \right)^n \right| \frac{\left(1 + |\xi_1|^2 \right)^{\gamma'/2}}{|\xi_1|^{\beta'}} \\
&= \frac{2^q \pi^{q-n}}{r^{q+1-n}} \frac{\left| \sin^n \left(\frac{\pi}{r} \xi_1 \right) \right| \left(1 + |\xi_1|^2 \right)^{\gamma'/2}}{|\xi_1|^{\beta'+n-q}}.
\end{aligned}
$$

To estimate an upper bound for $d_{\hat{\psi}_1}(\xi_1)$, we examine its behavior in three areas of ξ_1, i.e. $|\xi_1|$ tending to 0, to ∞ and the area in between. We show that we have upper bounds for all of these areas. Obviously, for any value range $[a, b] \in \mathbb{R}$ with $a > 0$ and $b < \infty$ we have an upper bound for $d_{\hat{\psi}_1}(\xi_1)$ since it is continuous.

For $|\xi_1| \to \infty$, we have

$$
d_{\hat{\psi}_1}(\xi_1) < \frac{\frac{2^{q+\gamma'/2}\pi^{q-n}}{r^{q+1-n}} \left| \sin^n \left(\frac{\pi}{r} \xi_1 \right) \right| |\xi_1|^{\gamma'}}{|\xi_1|^{\beta'+n-q}} \le K_1^\infty
$$

with $K_1^\infty > 0$ if $\gamma' \le \beta' + n - q$ since $\sin^n (\pi \xi_1)$ is bounded.

For $|\xi_1| \to 0$, we have that

$$
d_{\hat{\psi}_1}(\xi_1) < \frac{\frac{2^{q+\gamma'/2}\pi^{q-n}}{r^{p+1-n}} \left| \sin^n (\pi \xi_1) \right|}{|\xi_1|^{\beta'+n-q}} \le K_1^0
$$

if $p \ge \beta'$ since $p = n + 1$, while $K_1^0 > 0$. Therefore we have found $K_1 = \max \left(K_1^\infty, K_1^0 \right)$ in order to fulfill (2.13).

Concerning (2.14), we set

$$
d_{\hat{\psi}_2}(\xi_2) := \left| \hat{\psi}_2(\xi_2) \right| \left(1 + |\xi_2|^2 \right)^{\gamma'/2}
$$

which yields

$$
d_{\hat{\psi}_2}(\xi_2) = |\operatorname{sinc}^n(\xi_2)| \left(1 + |\xi_2|^2 \right)^{\gamma'/2} = \frac{\left| \sin^n (\pi \xi_2) \right|}{|\pi \xi_2|^n} \left(1 + |\xi_2|^2 \right)^{\gamma'/2}.
$$

As before, we examine the behavior of $d_{\hat{\psi}_2}(\xi_2)$ in the abovementioned three areas of its domain of definition. Also for $d_{\hat{\psi}_2}(\xi_2)$, its continuity shows an upper bound for the area between 0 and ∞. For $|\xi_2| \to 0$, this term is obviously bounded without any conditions. For $|\xi_2| \to \infty$, we have

$$
d_{\hat{\psi}_2}(\xi_2) < 2^{\gamma'/2} \pi^{-n} \left| \sin^n (\pi \xi_2) \right| \frac{|\xi_2|^{\gamma'}}{|\xi_2|^n} \le K_2
$$

for $K_2 > 0$ if $n \ge \gamma'$. This condition is already ensured by the conditions on ψ_1.

Finally, the properties of the sinc function yield that the we have

$$
\left\{ \xi_1 \in \mathbb{R} \colon \left| \hat{\psi}_1(\xi_1) \right|^2 = 0 \right\} = \{0, \pm r, \pm 2r, \ldots\}
$$

and

$$
\left\{ \xi_2 \in \mathbb{R} \colon \left| \hat{\psi}_2(\xi_2) \right|^2 = 0 \right\} = \{\pm 1, \pm 2, \ldots\}.
$$

Since $r > 1$, conditions (2.15) and (2.16) are also fulfilled. $\square$

3.2   Shearlet System and Transform

Given the definition of local precision shearlets, we now define the associated shearlet system and transform. As described earlier, basic shearlet systems, i.e. not cone-adapted ones, do have a directional bias such that detection of an edge along the x_1 axis of an image would be only possible with $s \to \infty$. We follow the idea of cone-adapted shearlets to use vertical and horizontal shearlets and adjust it to our time domain setting. Therefore we define a shearlet $\tilde{\psi}$ with

$$\tilde{\psi}\left(x_1, x_2\right) := \psi\left(x_2, x_1\right). \tag{3.11}$$

In that way we are able to cover the whole frequency plane with shear parameters $|s| \leq \tan\left(\pi/4\right) = 1$. However, we do not aim to tile the frequency domain into disjoint cones as in the cone-adapted approach [75]. Moreover, we do not restrict ourselves to parabolic scaling. Instead we choose a scaling matrix $A_{a,\alpha}$ given by

$$A_{a,\alpha} = \left(\begin{array}{cc} a & 0 \\ 0 & a^\alpha \end{array}\right)$$

with $a \in (0, 1]$ and $\alpha \in [1/2, 1)$. According to Kutyniok and Labate [75], the value $\alpha = 1/2$ is needed to get optimally sparse approximations of certain image models. However, in this thesis we are not concerned with optimally sparse approximations but with an optimal setup for the task of pedestrian detection. The lower the value of α the more elongated are the shearlets at fine scales. Thus, the directional response is intensively concentrated to the corresponding shear parameter $s \in \mathbb{R}$. For the practical application for pedestrian detection we choose α significantly larger than $1/2$ to ensure that we cover all directions when considering only few shearlets at fine scales. As we will see in Section 5.5.2, the choice of $\alpha > 1/2$ has a significant positive effect on the quality of our pedestrian detection algorithm.

For considerations in the frequency domain, we furthermore define a scaling function $\phi \in L^2\left(\mathbb{R}^2\right)$ that is associated with low frequencies. We set

$$\phi\left(x_1, x_2\right) := \psi_2\left(x_1\right)\psi_2\left(x_2\right). \tag{3.12}$$

Accordingly, we define the following shearlet system for the continuous case.

Definition 3.15. For ψ, $\tilde{\psi}$, $\phi \in L^2\left(\mathbb{R}^2\right)$ the *continuous local precision shearlet system* is given by

$$LPSH\left(\Phi, \Psi, \tilde{\Psi}\right) := \Phi\left(\phi\right) \cup \Psi\left(\psi\right) \cup \tilde{\Psi}\left(\tilde{\psi}\right)$$

with

$$\Phi\left(\phi\right) \quad := \quad \left\{\phi_t = \phi\left(\cdot - t\right) : t \in \mathbb{R}^2\right\},$$

$$\Psi\left(\psi\right) \quad := \quad \left\{\psi_{a,s,t} = a^{-\frac{(\alpha+1)}{2}}\psi\left(A_{a,\alpha}^{-1}S_s^{-1}\left(\cdot - t\right)\right) : a \in (0, 1], |s| \leq 1 + a^\alpha, t \in \mathbb{R}^2\right\},$$

$$\tilde{\Psi}\left(\tilde{\psi}\right) \quad := \quad \left\{\tilde{\psi}_{a,s,t} = a^{-\frac{(\alpha+1)}{2}}\tilde{\psi}\left(\tilde{A}_{a,\alpha}^{-1}S_s^{-T}\left(\cdot - t\right)\right) : a \in (0, 1], |s| \leq 1 + a^\alpha, t \in \mathbb{R}^2\right\}$$

and $\tilde{A}_{a,\alpha} := \mathrm{diag}\left(a^\alpha, a\right)$.

For the transform associated with the local precision shearlet system, we formulate its definition analogously to [75].

Definition 3.16. For

$$\mathbb{S}_L := \left\{ (a, s, t) : a \in (0, 1], \, s \le 1 + a^\alpha, \, t \in \mathbb{R}^2 \right\}$$

and $\psi, \tilde\psi, \phi \in L^2(\mathbb{R}^2)$, the *continuous local precision shearlet transform* of $f \in L^2(\mathbb{R}^2)$ is given by

$$LPST_{\phi,\psi,\tilde\psi}(f)(a,s,t) := \left(\langle f, \phi_t \rangle, \langle f, \psi_{a,s,t} \rangle, \left\langle f, \tilde\psi_{a,s,t} \right\rangle \right) \tag{3.13}$$

with $(t', (a,s,t), (\tilde a, \tilde s, \tilde t)) \in \mathbb{R}^2 \times \mathbb{S}_L^2$.

For the discrete setting used for implementation we define the shearlet system as follows.

Definition 3.17. For $\psi, \tilde\psi, \phi \in L^2(\mathbb{R}^2)$ the *local precision shearlet system* is given by

$$\mathcal{LPSH}\left(\Phi, \Psi, \tilde\Psi \right) := \Phi\left(\phi\right) \cup \Psi\left(\psi\right) \cup \tilde\Psi\left(\tilde\psi\right)$$

where

$$\begin{aligned}
\Phi\left(\phi\right) &:= \left\{ \phi_m = \phi\left(\cdot - cm\right) : \ m \in \mathbb{Z}^2 \right\}, \\
\Psi\left(\psi\right) &:= \left\{ \psi_{j,k,m} = 2^{\frac{j(\alpha+1)}{2}} \psi\left(A_{2^{-j},\alpha}^{-1} S_{s_{j,k}}^{-1} \cdot - cm \right) : j \ge 0, |k| \le \tilde\eta_j, m \in \mathbb{Z}^2 \right\}, \\
\tilde\Psi\left(\tilde\psi\right) &:= \left\{ \tilde\psi_{j,k,m} = 2^{\frac{j(\alpha+1)}{2}} \tilde\psi\left(\tilde A_{2^{-j},\alpha}^{-1} S_{s_{j,k}}^{-T} \cdot - cm \right) : j \ge 0, |k| \le \tilde\eta_j, m \in \mathbb{Z}^2 \right\},
\end{aligned}$$

with $s_{j,k} := \tan\left(k\pi / \eta_j\right)$, $\eta_j \in 2\mathbb{N}$, $\tilde\eta_j := \lceil (\eta_j/2 - 1)/2 \rceil$ and $c > 0$.

The parameter η_j denotes the number of shearlets at scale j. The shear parameter k is defined such that the directions are uniformly distributed along the circle. The partitioning in horizontal and vertical shearlets as well as the discretization of the shearing parameter k is illustrated for a fixed scale j and $\eta_j = 6$ in Figure 3.5. For the special case of η_j divisible by 4, each diagonal of $\mathbb{R}^2$ is covered by two shearlets, e.g. $\psi_{j,-1,m}$ and $\tilde\psi_{j,1,m}$. In this case we omit the horizontal shearlets $\tilde\psi$.

Definition 3.18. For $\psi, \tilde\psi, \phi \in L^2(\mathbb{R}^2)$ and $j \ge 0$, $k \le \tilde\eta_j$, $m \in \mathbb{Z}^2$, the discrete shearlet transform of an image $f \in L^2(\mathbb{R}^2)$ is then given by

$$\mathcal{LPST}_{\phi,\psi,\tilde\psi}(f)(j,k,m) := \left(\langle f, \phi_m \rangle, \langle f, \psi_{j,k,m} \rangle, \left\langle f, \tilde\psi_{j,k,m} \right\rangle \right) \tag{3.14}$$

which we call the *local precision shearlet transform (LPST)*.

3.3 Frame Property

In this section, we examine the properties of local precision shearlets and the corresponding shearlet transform concerning the ability to provide a frame for $L^2(\mathbb{R}^2)$. Let $\psi \in L^2(\mathbb{R}^2)$ be a local precision shearlet. Similar to [61], we obtain for the Fourier transform of $\psi_{a,s,t}$

$$\hat\psi_{a,s,t}(\xi) \quad = \quad a^{-\frac{(\alpha+1)}{2}} e^{-2\pi i \langle \xi, t \rangle} \mathcal{F}\left(\psi\left(\begin{pmatrix} \frac{1}{a} & -\frac{s}{a} \\ 0 & \frac{1}{a^\alpha} \end{pmatrix} \cdot \right) \right)(\xi)$$

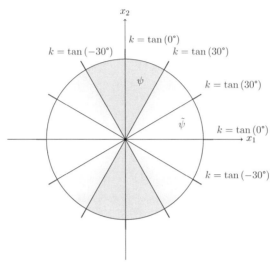

(a) Discretization of the shearing parameter k.

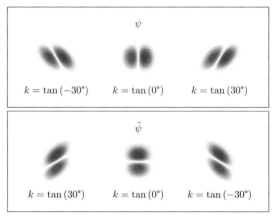

(b) Resulting vertical and horizontal spline shearlets (fine grid sampled) for $m = 5$.

Figure 3.5: Visualization of vertical and horizontal shearlets for a fixed scale j and $\eta_j = 6$.

$$= a^{-\frac{(\alpha+1)}{2}} e^{-2\pi i \langle \xi, t \rangle} \left(a^{-(\alpha+1)} \right)^{-1} \hat{\psi} \left(\begin{pmatrix} a & 0 \\ sa^\alpha & a^\alpha \end{pmatrix} \xi \right)$$

$$= a^{\frac{(\alpha+1)}{2}} e^{-2\pi i \langle \xi, t \rangle} \hat{\psi} \left(\begin{pmatrix} a & 0 \\ sa^\alpha & a^\alpha \end{pmatrix} \xi \right)$$

$$= a^{\frac{(\alpha+1)}{2}} e^{-2\pi i \langle \xi, t \rangle} \hat{\psi} \left(a\xi_1, a^\alpha \left(s\xi_1 + \xi_2 \right) \right).$$

With the separability condition (3.1) we have

$$\hat{\psi}_{a,s,t} (\xi) = a^{\frac{(\alpha+1)}{2}} e^{-2\pi i \langle \xi, t \rangle} \hat{\psi}_1 (a\xi_1) \, \hat{\psi}_2 \left(a^\alpha \left(s\xi_1 + \xi_2 \right) \right).$$

If ψ is a spline shearlet of order $p = n + 1$, $n \in \mathbb{N}$, and q-th derivative then according to Corollary 3.7 we get

$$\hat{\psi}_{a,s,t} (\xi) = a^{\frac{(\alpha+1)}{2}} e^{-2\pi i \langle \xi, t \rangle} \left(i2\pi\xi_1 \right)^q \operatorname{sinc}^n (a\xi_1) \operatorname{sinc}^n \left(a^\alpha \left(s\xi_1 + \xi_2 \right) \right).$$

Now, we analyze under which circumstances the local precision shearlet system forms a frame given a mother shearlet ψ that fulfills the conditions of Theorem 2.13. A main topic during this consideration is the required number of shearlets per scale such that the shearlet system covers the whole frequency domain. We set

$$\Theta := \left\{ (\xi_1, \xi_2) : \frac{1}{2} \leq |\xi_1| \leq 1, \, |\xi_2| \leq \frac{1}{2} \right\}.$$

According to conditions (2.13) - (2.16), we have $\Theta \subseteq \operatorname{ess\,supp} \hat{\psi}$. Therefore, Ψ is covering the frequency cones $\mathcal{C}_1 \cup \mathcal{C}_3$ if the scaled and sheared sets $\Theta_{j,k}$ with $j \geq 0$ and $|k| \leq \tilde{\eta}_j$ defined by

$$\begin{aligned} \Theta_{j,k} &:= A_{a_j,\alpha} S_{s_{j,k}}^T \Theta \\ &= \left\{ (\xi_1, \xi_2) : 1/2a_j \leq |\xi_1| \leq 1/a_j, \, |s_{j,k}\xi_1 + \xi_2| \leq 1/2a^\alpha \right\} \\ &= \left\{ (\xi_1, \xi_2) : 2^{j-1} \leq |\xi_1| \leq 2^j, \, |s_{j,k}\xi_1 + \xi_2| \leq 2^{j\alpha-1} \right\} \end{aligned}$$

are covering them. A set $\Theta_{j,k}$ is a trapezoid with $|\xi_1| \in \left[2^{j-1}, 2^j \right]$ and a lower boundary line $b_{j,k}^0$ which can be expressed by

$$\xi_2 = -s_{j,k}\xi_1 - 2^{j\alpha-1}$$

and an upper boundary line $b_{j,k}^1$ given by

$$\xi_2 = -s_{j,k}\xi_1 + 2^{j\alpha-1}.$$

For Ψ covering the frequency cones $\mathcal{C}_1 \cup \mathcal{C}_3$ we need

$$\bigcup_{|k| \leq \tilde{\eta}_j} \Theta_{j,k} = \left\{ (\xi_1, \xi_2) : 2^{j-1} \leq |\xi_1| \leq 2^j, \, |\xi_2/\xi_1| \leq 1 \right\},$$

for each $j \geq 0$. Similar considerations can be set up for $\tilde{\Psi}$ to cover the cones $\mathcal{C}_2 \cup \mathcal{C}_4$ while Φ clearly takes care of the low-frequency region $\mathcal{R}$.

Theorem 3.19. *Let $\psi \in L^2 \left(\mathbb{R}^2 \right)$ be a local precision shearlet that fulfills the conditions (2.13) - (2.16) with $\beta > 0$, $\gamma > 2 \left(\beta + 2 \right)$, $\beta' \geq \beta + \gamma$ and $\gamma' \geq \beta' - \beta + \gamma$. If for $\eta_j \in 2\mathbb{N}$, $j \geq 0$, and $\alpha \in [1/2, 1)$ we have*

$$\eta_j \geq \begin{cases} \dfrac{\pi}{2 \left(\operatorname{atan} \left(2^{j(\alpha-1)-1}+1 \right) - \frac{\pi}{4} \right)} & \textit{for } \operatorname{mod} (\eta_j, 4) \neq 0, \\[2ex] \dfrac{\pi}{\left(\pi/4 + \operatorname{atan} \left(2^{j(\alpha-1)} - 1 \right) \right)} & \textit{for } \operatorname{mod} (\eta_j, 4) = 0, \end{cases} \tag{3.15}$$

then there exists a $c_0 > 0$ such that the local precision shearlet system $\mathcal{LPSH} \left(\Phi, \Psi, \tilde{\Psi} \right)$ is a frame for $L^2 \left(\mathbb{R}^2 \right)$ for all $c \leq c_0$.

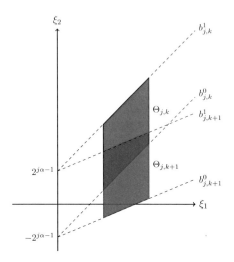

Figure 3.6: One side of the essential support of separable shearlets fulfilling conditions (2.13) - (2.16) for a fixed scale j and nearby shears.

Proof. We want to show that the conditions (2.11) and (2.12) adapted to our shearlet system are fulfilled. Similar to the proof of Corollary 2.17, see [86], we have in our case

$$
\begin{aligned}
\varphi\left(n, \mathcal{LPSH}\left(\Phi, \Psi, \tilde{\Psi}\right)\right) &= \varphi\left(n, \Phi\left(\phi\right) \cup \Psi\left(\psi\right) \cup \tilde{\Psi}\left(\tilde{\psi}\right)\right) \\
&= \operatorname*{ess\,sup}_{\xi \in \mathbb{R}^2} \sum_{j=0}^{\infty} \sum_{k=-\bar{\eta}_j}^{\bar{\eta}_j} \left|\hat{\psi}\left(A_{a_j, \alpha} S_{s_{j,k}}^T \xi\right)\right| \left|\hat{\psi}\left(A_{a_j, \alpha} S_{s_{j,k}}^T \xi + n\right)\right| \\
&\quad + \sum_{j=0}^{\infty} \sum_{k=-\bar{\eta}_j}^{k=\bar{\eta}_j} \left|\hat{\tilde{\psi}}\left(\tilde{A}_{a_j, \alpha} S_{s_{j,k}} \xi\right)\right| \left|\hat{\tilde{\psi}}\left(\tilde{A}_{a_j, \alpha} S_{s_{j,k}} \xi + n\right)\right| \\
&\quad + \left|\hat{\phi}\left(\xi\right)\right| \left|\hat{\phi}\left(\xi + n\right)\right| \\
&\leq \varphi\left(n, \Psi\left(\psi\right)\right) + \varphi\left(n, \tilde{\Psi}\left(\tilde{\psi}\right)\right) + \varphi\left(n, \Phi\left(\phi\right)\right)
\end{aligned}
$$

for (2.9) and

$$
\begin{aligned}
\tilde{\varphi}\left(\mathcal{LPSH}\left(\Phi, \Psi, \tilde{\Psi}\right)\right) &= \tilde{\varphi}\left(\Phi\left(\phi\right) \cup \Psi\left(\psi\right) \cup \tilde{\Psi}\left(\tilde{\psi}\right)\right) \\
&= \left|\hat{\phi}\left(\xi\right)\right|^2 + \sum_{j=0}^{\infty} \sum_{k=-\bar{\eta}_j}^{k=\bar{\eta}_j} \left|\hat{\psi}\left(A_{a_j, \alpha} S_{s_{j,k}}^T \xi\right)\right|^2 + \left|\hat{\tilde{\psi}}\left(\tilde{A}_{a_j, \alpha} S_{s_{j,k}} \xi\right)\right|^2 \\
&= \tilde{\varphi}\left(n, \Phi\left(\phi\right)\right) + \tilde{\varphi}\left(n, \Psi\left(\psi\right)\right) + \tilde{\varphi}\left(n, \tilde{\Psi}\left(\tilde{\psi}\right)\right)
\end{aligned}
$$

for (2.10).

The upper bound conditions, i.e. the existence of $D < \infty$ and the fulfillment of (2.12), follow from Theorem 3.14. Since there are upper bounds for a regular discrete shearlet system, we have upper bounds for each part of the local precision shearlet system. That means

$$
\tilde{\varphi}\left(\mathcal{LPSH}\left(\Phi,\Psi,\tilde{\Psi}\right)\right) \;\leq\; \left|\hat{\phi}\left(\xi\right)\right|^2 + 2 \sum_{j,k\in\mathbb{Z}} \left|\hat{\psi}\left(A_{a_j,\alpha} S^T_{s_{j,k}}\xi\right)\right|^2
$$
$$
\leq\; D_1 + 2D_2
$$
$$
:=\; D < \infty
$$

where $D_1 < \infty$ since $\hat{\phi}$ is obviously bounded. Concerning (2.12), we have $\varphi(n, \Psi(\psi)), \varphi(n, \tilde{\Psi}(\tilde{\psi})) \leq \varphi(n, SH(\psi, \Gamma_I))$ with $\varphi(n, SH(\psi, \Gamma_I))$ as in (2.9) for the setup of Theorem 2.13. Thus, for $c > 0$ we get

$$
\varphi\left(\frac{n}{c}, \mathcal{LPSH}\left(\Phi, \Psi, \tilde{\Psi}\right)\right) \leq 2\varphi\left(\frac{n}{c}, SH(\psi, \Gamma_I)\right) + \varphi\left(\frac{n}{c}, \Phi(\phi)\right).
$$

Next, we estimate $\varphi(\frac{n}{c}, \Phi(\phi))$. With $n = (n_1, n_2)$ we have

$$
\varphi\left(\frac{n}{c}, \Phi(\phi)\right) = \left|\hat{\phi}\left(\xi\right)\right|\left|\hat{\phi}\left(\xi + \frac{n}{c}\right)\right|
$$
$$
= \left|\hat{\psi}_2\left(\xi_1\right)\hat{\psi}_2\left(\xi_2\right)\right|\left|\hat{\psi}_2\left(\xi_1 + \frac{n_1}{c}\right)\hat{\psi}_2\left(\xi_2 + \frac{n_2}{c}\right)\right|
$$
$$
= \left|\hat{\psi}_2\left(\xi_1\right)\right|\left|\hat{\psi}_2\left(\xi_2\right)\right|\left|\hat{\psi}_2\left(\xi_1 + \frac{n_1}{c}\right)\right|\left|\hat{\psi}_2\left(\xi_2 + \frac{n_2}{c}\right)\right|
$$
$$
\leq K_1\left(1 + |\xi_1|^2\right)^{-\gamma'/2} K_2\left(1 + |\xi_2|^2\right)^{-\gamma'/2} \cdots
$$
$$
K_1\left(1 + \left|\xi_1 + \frac{n_1}{c}\right|^2\right)^{-\gamma'/2} K_2\left(1 + \left|\xi_2 + \frac{n_2}{c}\right|^2\right)^{-\gamma'/2},
$$

with $K_1, K_2 < \infty$ since ψ_2 fulfills condition (2.14). We set $\mu := \arg\max_{i=1,2}|n_i|$ and $\tilde{\mu} := \arg\min_{i=1,2}|n_i|$ such that we have $\|n\|_\infty = \max(|n_1|, |n_2|) = |n_\mu|$. We estimate the component in $\xi_{\tilde{\mu}}$-direction with $K_{\tilde{\mu}}(1 + |\xi_{\tilde{\mu}}|^2)^{-\gamma'/2} \leq K_{\tilde{\mu}}$. Then we have

$$
\varphi\left(\frac{n}{c}, \Phi(\phi)\right) \leq E_2\left(1 + |\xi_\mu|^2\right)^{-\gamma'/2}\left(1 + \left|\xi_\mu + \frac{n_\mu}{c}\right|^2\right)^{-\gamma'/2},
$$

with $E_2 < \infty$. Therefore, we have

$$
\varphi(\frac{n}{c}, \Phi(\phi)) \leq E_2 \min\left\{1, |\xi_\mu|^{-\gamma'}\right\} \min\left\{1, \left|\xi_\mu + \frac{n_\mu}{c}\right|^{-\gamma'}\right\}.
$$

For any $\xi, n \in \mathbb{R}^2$ we choose $c > 0$ such that $|n_\mu/2c| > \xi_\mu$. As described in [124], we then have $|\xi_\mu + n_\mu/c| \geq |n_\mu/c| - |\xi_\mu| \geq |n_\mu/2c|$ and thus $\max\{|\xi_\mu|, |\xi_\mu + n_\mu/c|\} \geq |n_\mu/2c|$. It follows that

$$
\varphi(\frac{n}{c}, \Phi(\phi)) \leq E_2 \left|\frac{n_\mu}{2c}\right|^{-\gamma'} = E_2\left(2c\right)^{\gamma'}\|n\|_\infty^{-\gamma'}.
$$

According to [86], we have

$$
\varphi\left(\frac{n}{c}, SH(\psi, \Gamma_I)\right) \leq E_3 c^{\gamma/2-\beta}\|n\|_\infty^{\beta-\gamma/2}
$$

with $E_3 < \infty$. It follows that

$$
\varphi\left(\frac{n}{c}, \mathcal{LPSH}\left(\Phi, \Psi, \tilde{\Psi}\right)\right) \leq 2E_3 c^{\gamma/2-\beta}\|n\|_\infty^{\beta-\gamma/2} + E_2\left(2c\right)^{\gamma'}\|n\|_\infty^{-\gamma'} \leq E_4 c^{\iota}\|n\|_\infty^{\iota'},
$$

with $E_4 < \infty$ and

$$\iota = \begin{cases} \gamma/2 - \beta & \text{for } c \geq 1 \\ \gamma' & \text{for } c < 1 \end{cases}, \quad \iota' = \begin{cases} \beta - \gamma/2 & \text{for } \|n\|_\infty \geq 1 \\ -\gamma' & \text{for } \|n\|_\infty < 1 \end{cases},$$

since $\beta > 0$, $\beta - \gamma/2 < -2$ and $\gamma' > \gamma > \beta$. Correspondingly, we get

$$\sum_{n \in \mathbb{Z}^2 \setminus \{0\}} \sqrt{\varphi\left(\frac{n}{c}, SH\left(\psi, \Gamma_I\right)\right) \varphi\left(-\frac{n}{c}, SH\left(\psi, \Gamma_I\right)\right)} \leq E_4 c^\iota \left(\sum_{m \in \mathbb{Z}^2 \setminus \{0\}} \|m\|_\infty^{\iota'}\right)$$

and according to [86], $\sum_{m \in \mathbb{Z}^2 \setminus \{0\}} \|m\|_\infty^{\beta - \gamma/2} < \infty$ when $\iota' < -2$. This condition is fulfilled since we assume $\beta > 0$, $\gamma > 2(\beta + 2)$, $\beta' \geq \beta + \gamma$ and $\gamma' \geq \beta' - \beta + \gamma$. Analogously as in [86], we conclude that for any $E > 0$, there exists a $c_0 > 0$ such that

$$\sum_{n \in \mathbb{Z}^2 \setminus \{0\}} \sqrt{\varphi\left(\frac{n}{c}, \mathcal{LPSH}\left(\Phi, \Psi, \tilde{\Psi}\right)\right) \varphi\left(-\frac{n}{c}, \mathcal{LPSH}\left(\Phi, \Psi, \tilde{\Psi}\right)\right)} < E$$

for all $c \leq c_0$.

For the lower bound, we show that the scaled and sheared sets $\Theta_{j,k}$ cover the frequency cones $\mathcal{C}_1 \cup \mathcal{C}_3$. For $\mathrm{mod}\,(\eta_j, 4) \neq 0$, we first need to show that the upper boundary line of the trapezoid $\Theta_{j,k}$ is reaching the diagonal for $|k| = (\eta_j/2 - 1)/2$. This condition can be expressed by

$$\left| -s_{j,k} \xi_1 + 2^{j\alpha - 1} \right| \geq |\xi_1|.$$

Because of the symmetry in possible values for k and due to the fact that $|k| < 1$, it is sufficient to examine this condition for $k = -(\eta_j/2 - 1)/2$ and $\xi_1 = 2^j$. We have

$$-s_{j,k} 2^j + 2^{j\alpha - 1} \geq 2^j,$$

$$-\tan\left(-\left(\frac{\eta_j}{4} - \frac{1}{2}\right)\frac{\pi}{\eta_j}\right) \geq 1 - 2^{j(\alpha-1)-1},$$

and therefore

$$\tan\left(\left(\frac{\eta_j}{4} - \frac{1}{2}\right)\frac{\pi}{\eta_j}\right) \geq 1 - 2^{j(\alpha-1)-1}. \tag{3.16}$$

We furthermore need to secure that no gaps between the trapezoids occur. That means the upper boundary line of $\Theta_{j,k+1}$ has to reach the lower line of $\Theta_{j,k}$. Therefore we have

$$-s_{j,k} 2^j - 2^{j\alpha - 1} \leq -s_{j,k+1} 2^j + 2^{j\alpha - 1}$$

and correspondingly

$$s_{j,k+1} - s_{j,k} \leq 2^{j(\alpha-1)}. \tag{3.17}$$

We set $\Delta_{s_{j,k}} := s_{j,k+1} - s_{j,k} = \tan\left((k+1)\frac{\pi}{\eta_j}\right) - \tan\left(k\frac{\pi}{\eta_j}\right)$, $k = -\tilde{\eta}_j, -\tilde{\eta}_j + \ldots \tilde{\eta}_j - 1$. Since $\tan$ is a strictly monotonically increasing function we have $\Delta_{s_{j,k}} \leq \Delta_{s_{j,\tilde{\eta}_j}}$. With (3.16) we get

$$\Delta_{s_{j,k}} \leq \tan\left(\left(\frac{\eta_j}{4} + \frac{1}{2}\right)\frac{\pi}{\eta_j}\right) - \tan\left(\left(\frac{\eta_j}{4} - \frac{1}{2}\right)\frac{\pi}{\eta_j}\right)$$

$$\leq \tan\left(\left(\frac{\eta_j}{4} + \frac{1}{2}\right)\frac{\pi}{\eta_j}\right) - \left(1 - 2^{j(\alpha-1)-1}\right)$$

$$= \tan\left(\left(\frac{\eta_j}{4} + \frac{1}{2}\right)\frac{\pi}{\eta_j}\right) + 2^{j(\alpha-1)-1} - 1.$$

Using this condition in (3.17) yields

$$\tan\left(\left(\frac{\eta_j}{4} + \frac{1}{2}\right)\frac{\pi}{\eta_j}\right) + 2^{j(\alpha-1)-1} - 1 \leq 2^{j(\alpha-1)},$$

$$\tan\left(\left(\frac{\eta_j}{4} + \frac{1}{2}\right)\frac{\pi}{\eta_j}\right) \leq 2^{j(\alpha-1)} - 2^{j(\alpha-1)-1} + 1$$

$$= 2^{j(\alpha-1)-1} + 1.$$

Solving this inequation for η_j provides

$$\left(\frac{\eta_j}{4} + \frac{1}{2}\right)\frac{\pi}{\eta_j} \leq \operatorname{atan}\left(2^{j(\alpha-1)-1} + 1\right),$$

$$\frac{\pi}{2\eta_j} \leq \operatorname{atan}\left(2^{j(\alpha-1)-1} + 1\right) - \frac{\pi}{4},$$

$$\eta_j \geq \frac{\pi}{2\left(\operatorname{atan}\left(2^{j(\alpha-1)-1} + 1\right) - \frac{\pi}{4}\right)}.$$

Under this condition, the cones $\mathcal{C}_2 \cup \mathcal{C}_4$ are covered by the sets $\widetilde{\Theta}_{j,k} := \widetilde{A}_{a_j,\alpha} S_{s_{j,k}} \widetilde{\Theta}$ with $\widetilde{\Theta} := \left\{(\xi_1, \xi_2) : |\xi_1| \leq \frac{1}{2}, \frac{1}{2} \leq |\xi_2| \leq 1\right\}$ analogously.

For the case of $\bmod (\eta_j, 4) = 0$, it is obvious that the diagonal is reached. In this case we need to ensure that the cones $\mathcal{C}_2 \cup \mathcal{C}_4$ are still covered completely although $\widetilde{\psi}_{j,-1,m}$ and $\widetilde{\psi}_{j,1,m}$ are omitted. Since we have

$$\widetilde{\Theta}_{j,k} = \left\{(\xi_1, \xi_2) : |\xi_1 + s_{j,k}\xi_2| \leq 2^{j\alpha-1}, \, 2^{j-1} \leq |\xi_2| \leq 2^j\right\}$$

we also have a right-hand side boundary line $\widetilde{b}_{j,k}^0$ with

$$\xi_2 = \frac{1}{s_{j,k}}\left(-\xi_1 + 2^{j\alpha-1}\right)$$

and a left-hand side boundary line $\widetilde{b}_{j,k}^1$ with

$$\xi_2 = \frac{1}{s_{j,k}}\left(-\xi_1 - 2^{j\alpha-1}\right).$$

Similar as before, we can examine under which condition $\widetilde{b}_{j,k}^0$ lies beneath $b_{j,-1}^1$ for $s_{j,k} = \tan\left(-(\eta_j/4 - 1)\pi/\eta_j\right) = \tan\left(-\pi/4 + \pi/\eta_j\right)$ and $\xi_1 = 2^j - 2^{j\alpha-1}$. The latter is the point of the diagonal at the boundary of $\widetilde{\Theta}_{j,k}$ in ξ_2 direction. We get

$$\frac{1}{s_{j,k}}\left(-2^j + 2^{j\alpha-1} + 2^{j\alpha-1}\right) \leq 2^j - 2^{j\alpha-1} + 2^{j\alpha-1},$$

$$s_{j,k} \leq \frac{-2^j + 2^{j\alpha}}{2^j},$$

and therefore

$$\tan\left(-\pi/4 + \pi/\eta_j\right) \leq 2^{j(\alpha-1)} - 1. \tag{3.18}$$

Also in this case, we need to ensure that (3.17) is fulfilled. We use the fact that $\Delta_{s_{j,k}} \leq \Delta_{s_{j,\tilde{\eta}_j - 1}}$ for all $k = -\tilde{\eta}_j, -\tilde{\eta}_j + \ldots \tilde{\eta}_j - 1$, i.e.

$$\Delta_{s_{j,k}} \leq \tan\left(\pi/4\right) - \tan\left(\pi/4 - \pi/\eta_j\right) = 1 - \tan\left(\pi/4 - \pi/\eta_j\right).$$

For (3.17) we get

$$1 - \tan\left(\pi/4 - \pi/\eta_j\right) \leq 2^{j(\alpha - 1)}$$

which is the same condition as (3.18). Solving for η_j yields

$$-\pi/4 + \pi/\eta_j \ \leq \ \operatorname{atan}\left(2^{j(\alpha-1)} - 1\right),$$

$$\eta_j \ \geq \ \frac{\pi}{\left(\pi/4 + \operatorname{atan}\left(2^{j(\alpha-1)} - 1\right)\right)}.$$

Obviously $\mathcal{R}$ is covered by Φ which finally yields

$$\left[-\frac{1}{2}, \frac{1}{2}\right]^2 \cup \left(\bigcup_{j \in \mathbb{Z}} \bigcup_{|k| \leq \tilde{\eta}_j} \Theta_{j,k} \cup \widetilde{\Theta}_{j,k}\right) = \mathbb{R}^2$$

and thus

$$\left|\hat{\phi}\left(\xi\right)\right|^2 + \sum_{j=0}^{\infty} \sum_{k=-\tilde{\eta}_j}^{k=\tilde{\eta}_j} \left|\hat{\psi}\left(A_{a_j,\alpha} S_{s_{j,k}}^T \xi\right)\right|^2 + \left|\hat{\tilde{\psi}}\left(\tilde{A}_{a_j,\alpha} S_{s_{j,k}} \xi\right)\right|^2$$

$$\geq K' \chi_{[-1/2, 1/2]^2}\left(\xi\right) + K\left(\sum_{j=0}^{\infty} \sum_{k=-\tilde{\eta}_j}^{k=\tilde{\eta}_j} \chi_{\Theta_{j,k}}\left(\xi\right) + \chi_{\widetilde{\Theta}_{j,k}}\left(\xi\right)\right)$$

$$> 0$$

for almost every $\xi \in \mathbb{R}^2$. $\qquad\qquad\qquad\qquad\qquad\qquad\qquad\qquad\qquad\qquad\qquad\qquad\qquad\square$

Corollary 3.20. *Let* $\psi \in L^2\left(\mathbb{R}^2\right)$ *be a spline shearlet of order* $p = n + 1$ *and q-th derivative with n and q as in Theorem 3.14. If η_j, $j \geq 0$, fulfills (3.15), then there exists a $c_0 > 0$ such that* $\mathcal{LPSH}\left(\Phi, \Psi, \widetilde{\Psi}\right)$ *is a frame for* $L^2\left(\mathbb{R}^2\right)$ *for all* $c \leq c_0$.

Given the result that our shearlet system forms a frame for $L^2(\mathbb{R}^2)$, it is possible to reconstruct a signal $f \in L^2(\mathbb{R}^2)$ by its shearlet coefficients. Although this topic is not in the focus of this thesis, we will evaluate in the next section which algorithms are applicable for a signal reconstruction with our shearlet system.

3.4   Signal Reconstruction

To reconstruct a signal $f \in L^2(\mathbb{R}^2)$, one can make use of the frame operator

$$S : L^2(\mathbb{R}^2) \to L^2(\mathbb{R}^2), \quad f \mapsto \sum_{i \in I} \langle f, \varphi_i \rangle \, \varphi_i$$

of a frame $\Phi = (\varphi_i)_{i \in I}$ for $L^2\left(\mathbb{R}^2\right)$. According to Christensen [17], a signal $f \in L^2\left(\mathbb{R}^2\right)$ can be reconstructed by the formula (2.2)

$$f = \sum \langle f, \varphi_i \rangle \, S^{-1} \varphi_i.$$

Therefore, we need to find the inverse frame operator S^{-1} associated to our shearlet frame $\mathcal{LPSH}(\Phi, \Psi, \tilde{\Psi})$ to provide an explicit reconstruction formula. As described in [17], the inversion of the frame operator can be very complicated in practice. As an alternative, the following algorithm, also known as the *frame algorithm*, is provided.

Lemma 3.21 ([17]). *Let $\Phi = (\varphi_i)_{i \in I}$ be a frame for $L^2(\mathbb{R}^2)$ with frame bounds A and B. For $f \in L^2(\mathbb{R}^2)$, we define functions $(g_i)_{i=0}^{\infty}$ in $L^2(\mathbb{R}^2)$ by*

$$g_0 = 0, \; g_i = g_{i-1} + \frac{2}{A+B} S \left(f - g_{i-1} \right), \; i \geq 1. \tag{3.19}$$

Then

$$\|f - g_i\| \leq \left(\frac{B-A}{B+A} \right)^i \|f\|.$$

The sequence elements g_i in (3.19) converge to f for $i \to \infty$, while their computation depends on the frame bounds A and B. First, we need to know the frame bounds in order to apply the frame algorithm. Second, the ratio of the frame bounds determines the speed of convergence to f. It follows that g_i might only converge slowly to f in case B is much larger than A. This could refer to the cause that either the estimate of the frame bounds is not optimal or the frame is far from being tight. In order to get to a faster convergence, Gröchenig [45] applies the *Chebyshev method* and the *conjugate gradients method*.

Theorem 3.22 (Chebyshev algorithm [45]). *Let $\Phi = (\varphi_i)_{i \in I}$ be a frame for $L^2 (\mathbb{R}^2)$ with frame bounds A, B and let*

$$\rho := \frac{B-A}{B+A}, \; \sigma := \frac{\sqrt{B} - \sqrt{A}}{\sqrt{B} + \sqrt{A}}.$$

For $f \in L^2 (\mathbb{R}^2)$, we define functions $(g_i)_{i=0}^{\infty}$ in $L^2 (\mathbb{R}^2)$ and corresponding numbers $(\lambda_i)_{i=1}^{\infty}$ by

$$g_0 = 0, \; g_1 = \frac{2}{A+B} Sf, \; \lambda_1 = 2,$$

and for $i \geq 2$

$$\lambda_i = \frac{1}{1 - \frac{\rho^2}{4} \lambda_{i-1}}, \; g_i = \lambda_i \left(g_{i-1} - g_{i-2} + \frac{2}{A+B} S \left(f - g_{i-1} \right) \right) + g_{i-2}.$$

Then

$$\|f - g_i\| \leq \frac{2\sigma^i}{1 + \sigma^{2i}} \|f\|.$$

Although, the elements g_i converge faster to f in the Chebyshev algorithm than in the frame algorithm, still the frame bounds are needed. This is not the case for the conjugate gradients algorithm.

Theorem 3.23 (Conjugate gradient algorithm [45]). *Let $\Phi = (\varphi_i)_{i \in I}$ be a frame for $L^2 (\mathbb{R}^2)$ and let $f \in L^2(\mathbb{R}^2) \backslash f_0$, where f_0 is the zero function. We define functions $(g_i)_{i=0}^{\infty}$, $(r_i)_{i=0}^{\infty}$, $(p_i)_{i=0}^{\infty}$ and numbers $(\lambda_i)_{i=0}^{\infty}$ by*

$$g_0 = 0, \; r_0 = p_0 = Sf, \; p_{-1} = 0$$

and, for $i \geq 2$,

$$\lambda_i = \frac{\langle r_i, p_i \rangle}{\langle p_i, Sp_i \rangle},$$

$$g_{i+1} = g_i + \lambda_i p_i,$$
$$r_{i+1} = r_i - \lambda_i S p_i$$
$$p_{i+1} = S p_i - \frac{\langle S p_i, S p_i \rangle}{\langle p_i, S p_i \rangle} p_i - \frac{\langle S p_i, S p_{i-1} \rangle}{\langle p_{i-1}, S p_{i-1} \rangle} p_{i-1}.$$

Then $g_i \to f$ for $i \to \infty$.

According to Theorem 2.12, the frame bounds A and B for our shearlet frame $\mathcal{LPSH}(\Phi, \Psi, \widetilde{\Psi})$ can be estimated by $1/c^2 (C - E) \leq A \leq B \leq 1/c^2 (D + E)$. Unfortunately, the constants $c, C, D, E < \infty$ are not known explicitly. Therefore, the only possibility to approximate a function $f \in L^2(\mathbb{R}^2)$ by one of the presented algorithms is the conjugate gradients algorithm. However, since signal reconstruction is not the main concern in this thesis, we do not realize an application of this algorithm and an analysis of the corresponding results.

3.5   Practical Application

The application of our shearlet design is crucial for the quality of pedestrian detection. It gives us the ability to define shearlets compactly supported in time domain of any size such that only the nearest neighborhood of m is considered for computing the shearlet transform at location m and fine scales. This fact leads to very precise edge detections in comparison to other known shearlets. See Section 5.5 for more details. We compute our shearlet filters by sampling shearlets directly in time domain with a sampling constant $c > 0$. Additionally, in the style of the *Fast Finite Shearlet Transform (FFST)* [61], we consider digital images in $\mathbb{R}^{M \times N}$ as functions $f \in L^2(\mathbb{R}^2)$ sampled on a grid $\mathcal{G}$. In our case we define this grid by

$$\mathcal{G} := \{(cm_1, cm_2): m_1 = -\lfloor M/2 \rfloor, \ldots, \lceil M/2 \rceil - 1,$$
$$m_2 = -\lfloor N/2 \rfloor, \ldots, \lceil N/2 \rceil - 1\}.$$

Finally, the LPST is computed directly in time domain using the 2D convolution $f * \psi_{j,k,0}$.

Now, we briefly address the topic of computational complexity. In contrast to our approach, shearlet transforms are usually calculated by applying the 2D Fast Fourier transform (FFT) and its inverse (IFFT). As described by Duval-Poo et al. [34] the computational complexity of the shearlet transform of a $N \times N$ sized image using FFT and IFFT is $\mathcal{O}(j_0 \eta N^2 + j_0 \eta N^2 \log(N))$, $\eta = \sum_{j=0}^{j_0-1} \eta_j$. The parameter $j_0 \in \mathbb{N}$ describes the number of scales considered during computation. This computational complexity can be reduced to $\mathcal{O}(N^2 \log(N))$ since the number of all shearlets $j_0 \eta$ can be assumed as a small constant compared to N. According to the authors, using a 2D convolution with $W \times W$ sized shearlets results in $\mathcal{O}(j_0 \eta N^2 W^2)$ that reduces to $\mathcal{O}(N^2)$ if $W \ll N$. Using local precision shearlets, the condition $W \ll N$ can be easily fulfilled which leads to a significantly reduced computational complexity. When using small sized shearlets, e.g. 16×16, we observe a decrease by an order of magnitude in the runtime for computing the shearlet transform of a 640×480 image.

3.6   Conclusion

In this chapter, we designed our own mother shearlets and shearlet system in order to have the capability to compute highly qualitative image features. Concerning the design of mother shearlets, we used compactly supported, separable functions with a point-symmetric wavelet in the first and an axis-symmetric bump function in the second component. As an important example of our shearlets, we introduced spline shearlets. Such a shearlet ψ consists of a derivative of a B-spline in the ψ_1 and the corresponding B-spline itself in the ψ_2 component. We showed that spline shearlets are admissible and that the continuous shearlet transform is a multiple of an isometry with their usage. In addition, we showed that they are suitable components for a regular discrete shearlet system to form a frame for $L^2(\mathbb{R}^2)$.

Furthermore, we defined our discrete shearlet system with evenly distributed orientations of the involved shearlets and a high degree of flexibility concerning the number of shearlets per scale. We showed that this shearlet system forms a frame for $L^2(\mathbb{R}^2)$ provided that the mother functions fulfill sufficient conditions and that we use enough shearlets per scale. We derived a required number of shearlets per scale depending on the degree of anisotropy α.

Although the topic of signal reconstruction is not in the focus of this thesis, we examined which algorithms could be used to reconstruct a signal $f \in L^2(\mathbb{R}^2)$ given its frame coefficients. We concluded that the conjugate gradients algorithm is the only possibility to reconstruct a signal with our shearlet system.

Finally, we discussed the practical application of our shearlets. We described that our digital shearlet filters are obtained by a sampling directly in time domain. With a suitable sampling of the input image, our shearlet transform is computed by a convolution in time domain, which yields benefits concerning the computational costs of it.

"There are things known and there are things unknown,
and in between are the doors of perception."

Aldous Huxley

4

Edge Detection using Local Precision Shearlets

In this chapter, we study the characterization of edges based on the properties of the continuous shearlet transform $\mathcal{SH}_\psi$ using a local precision shearlet ψ. The precise detection of edges is the key factor for the extraction of highly qualitative image feature using shearlets. It is the basis for a beneficial application of shearlets either in hand-crafted feature detectors or in CNN algorithms. We study the ability of local precision shearlets to characterize edge points in the manner of Guo and Labate [53]. The image function f is modeled as the characteristic function χ_R of a bounded domain $R \subset \mathbb{R}^2$ with piecewise smooth boundary ∂R. We will use the notation of [53] already presented in Section 2.4. This model scenario has been analyzed in several papers for the case of band-limited shearlets [50, 53, 54]. In this case, the boundary ∂R, the orientation and the type of the edge point can be deduced by the decay rate of the continuous shearlet transform, see 2.19. As indicated before, band-limited shearlets have infinite support in the space domain. The spatial localization of compactly supported shearlets can lead to improved edge classification in comparison to band-limited shearlets. Kutyniok and Petersen [78] analyzed the scenario if one replaces band-limited by compactly supported shearlets. The results are presented in 2.21.

Although the results in [78] resolve the issues using band-limited shearlets described in Section 2.4, certain additional assumptions on the compactly supported shearlets lead to drawbacks in a practical application. First of all, the separable shearlet $\psi = \psi_1\psi_2$ shall fulfill in its second function component ψ_2 the condition $\psi_2'(0) \neq 0$ while ψ_1 shall be a compactly supported wavelet. This condition induces an asymmetry of the shearlet with respect to x_2 that leads to a stronger response of the shearlet transform on one side of an edge. Second, some results require that the shearlet possesses a minimum number of vanishing moments. As indicated before, increasing the vanishing moments increases the oscillations of a shearlet which create artifacts in the edge detection result reducing the detection precision. As we will see in 4.2, a local precision shearlet with just one vanishing moment creates visibly the best edge detection results. More importantly, we will show in 5.5.2 that the best pedestrian detection results are achieved with just one vanishing moment. Fittingly, the theoretical results we derive in this chapter for the characterization of edge points using local precision shearlets also require a condition that limits the number of vanishing moments to one.

In contrast to the results from the publications stated above, we use the more flexible scaling matrix $A_{a,\alpha}$ with degree of anisotropy $\alpha \in [1/2, 1)$. We derive decay rates depending on α, which make it possible to detect edge points and the orientation of the corresponding edge. In case of the special case $\alpha = 1/2$, regular points and corner points of the first and second type can be distinguished from one another by the limit value of the shearlet transform for $a \to 0$ in case s corresponds to the normal direction at the analyzed point $p \in \partial R$. Remarkably, in this case the symmetry of local precision shearlets improve the results of 2.21, originally described in [78], concerning this limit value.

4.1   Characterization of Edge Points

First of all, for a local precision shearlet ψ, points p outside of the boundary do not have to be characterized by the asymptotic decay of $\mathcal{SH}_\psi \chi_R (a, s, p)$ for $a \to 0+$. For such points $p \notin \partial R$, we derive the following statement.

Proposition 4.1. *Let $\psi \in L^2 (\mathbb{R}^2)$ be a local precision shearlet and $R \subset \mathbb{R}^2$ with boundary ∂R of length L to be smooth except for finitely many corner points. For $p \notin \partial R$ and each $s \in \mathbb{R}$, there is a scale $a_0(s) \in \mathbb{R}$ small enough such that we have $\mathcal{SH}_\psi \chi_R (a_0(s), s, p) = 0$.*

Proof. We analyze

$$
\begin{aligned}
\mathcal{SH}_\psi \chi_R (a, s, p) &= \langle \chi_R, \psi_{a,s,p} \rangle \\
&= \int_{\mathbb{R}^2} \chi_R (x)\, \psi_{a,s,p} (x)\, \mathrm{d}x \\
&= \int_R \psi_{a,s,p} (x)\, \mathrm{d}x \\
&= \int_{R \cap \mathrm{supp}\, \psi_{a,s,p}} \psi_{a,s,p} (x)\, \mathrm{d}x.
\end{aligned}
$$

Without loss of generality, we can assume $p = (0,0)$ and R being such that $p \notin \partial R$. For all other cases $p' \neq (0,0)$, we just need to shift ψ and R by p' which results in the same integration result. Therefore, we examine

$$
\mathcal{SH}_\psi \chi_R (a, s, p) = \int_{R \cap \mathrm{supp}\, \psi_{a,s,p}} \psi_{a,s,0} (x)\, \mathrm{d}x.
$$

Since

$$
\begin{aligned}
\psi_{a,s,0} (x) &= a^{-\frac{1+\alpha}{2}} \psi \left(A_{a,\alpha}^{-1} S_s^{-1} x \right) \\
&= a^{-\frac{1+\alpha}{2}} \psi \left(\begin{pmatrix} \frac{1}{a} & -\frac{s}{a} \\ 0 & \frac{1}{a^\alpha} \end{pmatrix} \begin{pmatrix} x_1 \\ x_2 \end{pmatrix} \right) \\
&= a^{-\frac{1+\alpha}{2}} \psi_1 \left(\frac{1}{a} x_1 - \frac{s}{a} x_2 \right) \psi_2 \left(\frac{1}{a^\alpha} x_2 \right)
\end{aligned}
$$

as well as $\mathrm{supp}\, \psi_1 \subseteq [-b_1, b_1]$ and $\mathrm{supp}\, \psi_2 \subseteq [-b_2, b_2]$ with $b_1, b_2 \in \mathbb{R}^+$, we have

$$
\mathrm{supp}\, \psi_{a,s,0} \subseteq \{ (x_1, x_2) : \ |x_1| \le ab_1 + sa^\alpha b_2, \ |x_2| \le a^\alpha b_2 \}.
$$

For $p \notin R$, there is an $a_0(s) \in \mathbb{R}^+$ small enough such that $R \cap \mathrm{supp}\, \psi_{a,s,0} = \emptyset$ and therefore

$$
\mathcal{SH}_\psi \chi_R (a, s, 0) = \int_{R \cap \mathrm{supp}\, \psi_{a,s,0}} \psi_{a,s,0} (x)\, \mathrm{d}x
$$

$$= \int_{\varnothing} \psi_{a,s,0}(x)\,\mathrm{d}x$$
$$= 0.$$

For the remaining possibility $p \in R$, there exists an $a_0(s) \in \mathbb{R}^+$ small enough such that $R \cap \mathrm{supp}\,\psi_{a,s,0} = \mathrm{supp}\,\psi_{a,s,0}$. In this case we get

$$
\begin{aligned}
\mathcal{SH}_{\psi}\chi_R(a,s,0) &= \int_{\mathbb{R}^2} \psi_{a,s,0}(x)\,\mathrm{d}x \\
&= \int_{\mathbb{R}^2} a^{-\frac{1+\alpha}{2}} \psi\left(A_{a,\alpha}^{-1}S_s^{-1}x\right)\mathrm{d}x.
\end{aligned}
$$

With $\sigma_{A_{a,\alpha},S_s}(x) := A_{a,\alpha}^{-1}S_s^{-1}x$, we rewrite this equation by

$$
\mathcal{SH}_{\psi}\chi_R(a,s,0) = a^{\frac{1+\alpha}{2}} \int_{\mathbb{R}^2} \det\left(\nabla\sigma_{A_{a,\alpha},S_s}\right)\psi\left(\sigma_{A_{a,\alpha},S_s}(x)\right)\mathrm{d}x.
$$

By applying the transformation theorem and by the separability of ψ we get

$$
\begin{aligned}
\mathcal{SH}_{\psi}\chi_R(a,s,0) &= a^{\frac{1+\alpha}{2}} \int_{\mathbb{R}^2} \psi(x)\,\mathrm{d}x \\
&= a^{\frac{1+\alpha}{2}} \int_{\mathbb{R}} \psi_1(x_1)\,\mathrm{d}x_1 \int_{\mathbb{R}} \psi_2(x_2)\,\mathrm{d}x_2.
\end{aligned}
$$

The component ψ_1 is defined as wavelet, i.e. $\int_{\mathbb{R}} \psi_1(x_1)\mathrm{d}x_1 = 0$. This fact yields

$$
\begin{aligned}
\mathcal{SH}_{\psi}\chi_R(a,s,0) &= a^{\frac{1+\alpha}{2}} \underbrace{\int_{\mathbb{R}} \psi_1(x_1)\,\mathrm{d}x_1}_{=0} \int_{\mathbb{R}} \psi_2(x_2)\,\mathrm{d}x_2 \\
&= 0.
\end{aligned}
$$

$\square$

Regarding practical application, especially for detecting points of a pedestrian's silhouette, regular points of ∂R are of major interest. In case that the shear parameter s does not correspond to the normal direction of ∂R at p, we make use of the special properties of a local precision shearlet to state a specific decay rate of the shearlet transform. Theorem 2.21 only provides an estimation depending on the number of vanishing moments and the differentiability of the shearlet in this case. In the opposite case that s corresponds the normal direction, we derive a decay rate that matches the one of Theorem 2.21 for $\alpha = 1/2$. For this special case, we have the same limit value of the shearlet transform as in Theorem 2.21.

Theorem 4.2. *Let $\psi \in L^2(\mathbb{R}^2)$ be a local precision shearlet with $\mathrm{supp}\,\psi \subseteq [-b_1,b_1] \times [-b_2,b_2] \subset \mathbb{R}^2$, $\int_{-b_1}^{0} \psi_1(x_1)\,x_1\mathrm{d}x_1 \neq 0$ and $\psi_2 \in C^2(\mathbb{R}) \cap L^2(\mathbb{R})$ as well as $R \subset \mathbb{R}^2$ with boundary ∂R of length L to be smooth except for finitely many corner points. Let furthermore $p = \vec{\alpha}(t_0)$, $t_0 \in (0,L)$, be a regular point of ∂R.*

> *i. If $s = s_0$ does not correspond to the normal direction of ∂R at p, then*
>
> $$\lim_{a \to 0+} a^{-\frac{3-\alpha}{2}} |\mathcal{SH}_{\psi}\chi_R(a,s,p)| > 0.$$

ii. If $s = s_0$ corresponds to the normal direction of ∂R at p, then

$$\lim_{a \to 0+} a^{-\frac{1+\alpha}{2}} \left| \mathcal{SH}_\psi \chi_R (a,s,p) \right| > 0.$$

If furthermore we have $\alpha = 1/2$, then

$$\lim_{a \to 0+} a^{-\frac{3}{4}} \mathcal{SH}_\psi \chi_R (a,s,p) = \int_S \psi(x) \, \mathrm{d}x \quad \text{if } s \in B_a(s_0),$$

with $B_a(s_0) = \{s \in \mathbb{R} \colon |s - s_0| \le a\}$,

$$S := \left\{ (x_1, x_2) \in \operatorname{supp} \psi \colon x_1 \le \frac{\bar{\alpha}_1''(t_0) - s \bar{\alpha}_2''(t_0)}{2\rho(s)^2} x_2^2 \right\}$$

and $\rho(s) := \cos(\operatorname{atan}(s))$.

Proof. The proof is separated according to the two items of the theorem.

i. According to the proof of Theorem 2.21 [78], we can analyze the situation for $\langle \chi_R, \psi_{a,0,0} \rangle$ since the result can directly be transferred to the situation of a general s by considering $f := \chi_R \circ S_s$ instead of χ_R. According to Kutyniok and Petersen [78], $\bar{\alpha}$ is locally given as the graph of a function $g \colon [-\varepsilon, \varepsilon] \to [-\varepsilon, \varepsilon]$, since it is differentiable at 0 and its normal does not equal $\pm(1, 0)$, such that for a small enough

$$\langle \chi_R, \psi_{a,0,0} \rangle = \int_{x_2 \ge g(x_1)} \psi_{a,0,0}(x) \, \mathrm{d}x.$$

Since $g(0) = 0$ and $\alpha(t_0)$ is a regular point of ∂R, i.e. infinitely many times differentiable at $t_0 = 0$, a Taylor expansion of g at 0 provides

$$\langle \chi_R, \psi_{a,0,0} \rangle = \int_{x_2 \ge g'(0)x_1 + R_1 g(x_1, 0)} \psi_{a,0,0}(x) \, \mathrm{d}x$$

with the remainder term $R_1 g(x_1, 0)$ of the Taylor expansion at 0. We rewrite the last formula by

$$
\begin{aligned}
\langle \chi_R, \psi_{a,0,0} \rangle &= \int_T a^{-\frac{1+\alpha}{2}} \psi \left(A_a^{-1} x \right) \mathrm{d}x \\
&= \int_T a^{-\frac{1+\alpha}{2}} \psi \left(\sigma_{A_{a,\alpha}}(x) \right) \mathrm{d}x \\
&= a^{\frac{1+\alpha}{2}} \int_T \det \left(\nabla \sigma_{A_{a,\alpha}}(x) \right) \psi \left(\sigma_{A_{a,\alpha}}(x) \right) \mathrm{d}x
\end{aligned}
$$

with $\sigma_{A_{a,\alpha}}(x) := A_{a,\alpha}^{-1} x$ and $T := \{(x_1, x_2) \in \mathbb{R}^2 \colon x_2 \ge g'(0) x_1 + R_1 g(x_1, 0)\}$. An application of the transformation theorem yields

$$
\begin{aligned}
\langle \chi_R, \psi_{a,0,0} \rangle &= a^{\frac{1+\alpha}{2}} \int_{\sigma_{A_{a,\alpha}}(T)} \psi(x) \, \mathrm{d}x \\
&= a^{\frac{1+\alpha}{2}} \int_{A_{a,\alpha}^{-1} T} \psi(x) \, \mathrm{d}x \\
&= a^{\frac{1+\alpha}{2}} \int_{a^\alpha x_2 \ge g'(0) a x_1 + R_1 g(a x_1, 0)} \psi(x) \, \mathrm{d}x.
\end{aligned}
$$

Since $\alpha(t_0)$ is a regular point of ∂R, we can use the *Lagrange form* [70] of the remainder $R_1 g(x_1, 0)$ and get

$$R_1 g(ax_1, 0) = \frac{1}{2} g''(\zeta) a^2 x_1^2,$$

for $0 < \zeta < ax_1$. With the separability of local precision shearlets, we have

$$
\begin{aligned}
\langle \chi_R, \psi_{a,0,0} \rangle &= a^{\frac{1+\alpha}{2}} \int_{a^\alpha x_2 \geq g'(0) a x_1 + \frac{1}{2} g''(\zeta) a^2 x_1^2} \psi(x)\, \mathrm{d}x \\
&= a^{\frac{1+\alpha}{2}} \int_{x_2 \geq g'(0) a^{1-\alpha} x_1 + \frac{1}{2} g''(\zeta) a^{2-\alpha} x_1^2} \psi_1(x_1)\, \psi_2(x_2)\, \mathrm{d}x.
\end{aligned}
$$

We set $t(x_1) := g'(0) a^{1-\alpha} x_1 + \frac{1}{2} g''(\zeta) a^{2-\alpha} x_1^2$ and split the integral

$$
\begin{aligned}
\langle \chi_R, \psi_{a,0,0} \rangle &= a^{\frac{1+\alpha}{2}} \int_{x_2 \geq t(x_1)} \psi_1(x_1)\, \psi_2(x_2)\, \mathrm{d}x \\
&= a^{\frac{1+\alpha}{2}} \int_{\substack{t(x_1) \leq 0 \\ x_2 \geq t(x_1)}} \psi_1(x_1)\, \psi_2(x_2)\, \mathrm{d}x + a^{\frac{1+\alpha}{2}} \int_{\substack{t(x_1) \geq 0 \\ x_2 \geq t(x_1)}} \psi_1(x_1)\, \psi_2(x_2)\, \mathrm{d}x.
\end{aligned}
$$

In that way, the first integral takes care about negative values of $t(x_1)$ and the second one deals with positive values of $t(x_1)$. Thus, we have

$$
\begin{aligned}
\langle \chi_R, \psi_{a,0,0} \rangle &= a^{\frac{1+\alpha}{2}} \int_{0 \geq x_2 \geq t(x_1)} \psi_1(x_1)\, \psi_2(x_2)\, \mathrm{d}x + a^{\frac{1+\alpha}{2}} \int_{\substack{t(x_1) \leq 0 \\ x_2 \geq 0}} \psi_1(x_1)\, \psi_2(x_2)\, \mathrm{d}x \\
&\quad + a^{\frac{1+\alpha}{2}} \int_{x_2 \geq t(x_1) \geq 0} \psi_1(x_1)\, \psi_2(x_2)\, \mathrm{d}x \\
&= a^{\frac{1+\alpha}{2}} \int_{0 \geq x_2 \geq t(x_1)} \psi_1(x_1)\, \psi_2(x_2)\, \mathrm{d}x + a^{\frac{1+\alpha}{2}} \int_{\substack{t(x_1) \leq 0 \\ x_2 \geq 0}} \psi_1(x_1)\, \psi_2(x_2)\, \mathrm{d}x \\
&\quad + a^{\frac{1+\alpha}{2}} \int_{\substack{t(x_1) \geq 0 \\ x_2 \geq 0}} \psi_1(x_1)\, \psi_2(x_2)\, \mathrm{d}x - a^{\frac{1+\alpha}{2}} \int_{0 \leq x_2 \leq t(x_1)} \psi_1(x_1)\, \psi_2(x_2)\, \mathrm{d}x \\
&= a^{\frac{1+\alpha}{2}} \int_{0 \geq x_2 \geq t(x_1)} \psi_1(x_1)\, \psi_2(x_2)\, \mathrm{d}x + a^{\frac{1+\alpha}{2}} \int_{x_2 \geq 0} \psi_1(x_1)\, \psi_2(x_2)\, \mathrm{d}x \\
&\quad - a^{\frac{1+\alpha}{2}} \int_{0 \leq x_2 \leq t(x_1)} \psi_1(x_1)\, \psi_2(x_2)\, \mathrm{d}x.
\end{aligned}
$$

Since ψ integrates to 0 along x_1, we have $a^{\frac{1+\alpha}{2}} \int_{x_2 \geq 0} \psi_1(x_1)\, \psi_2(x_2)\, \mathrm{d}x = 0$, which results in

$$
\langle \chi_R, \psi_{a,0,0} \rangle = a^{\frac{1+\alpha}{2}} \int_{0 \geq x_2 \geq t(x_1)} \psi_1(x_1)\, \psi_2(x_2)\, \mathrm{d}x \tag{4.1}
$$

$$
- a^{\frac{1+\alpha}{2}} \int_{0 \leq x_2 \leq t(x_1)} \psi_1(x_1)\, \psi_2(x_2)\, \mathrm{d}x. \tag{4.2}
$$

Next, we apply a Taylor expansion of ψ_2 at 0 and use the fact that for a local precision shearlet we have $\psi_2'(0) = 0$. For (4.1), we get

$$
\begin{aligned}
&a^{\frac{1+\alpha}{2}} \int_{0 \geq x_2 \geq t(x_1)} \psi_1(x_1)\, \psi_2(x_2)\, \mathrm{d}x \\
&= a^{\frac{1+\alpha}{2}} \int_{-b_1}^{b_1} \int_{t(x_1)}^{0} \psi_1(x_1)\, \psi_2(x_2)\, \mathrm{d}x_2 \mathrm{d}x_1 \\
&= a^{\frac{1+\alpha}{2}} \int_{-b_1}^{b_1} \int_{t(x_1)}^{0} \psi_1(x_1)\, (\psi_2(0) + \underbrace{\psi_2'(0)}_{=0} x_2 + R_1 \psi_2(x_2, 0))\mathrm{d}x_2 \mathrm{d}x_1
\end{aligned}
$$

$$= a^{\frac{1+\alpha}{2}} \int_{-b_1}^{b_1} \int_{t(x_1)}^{0} \psi_1(x_1)\left(\psi_2(0) + R_1\psi_2(x_2, 0)\right) \mathrm{d}x_2 \mathrm{d}x_1$$

$$= a^{\frac{1+\alpha}{2}} \int_{-b_1}^{b_1} \int_{t(x_1)}^{0} \psi_1(x_1)\psi_2(0)\,\mathrm{d}x_2\mathrm{d}x_1 + a^{\frac{1+\alpha}{2}} \int_{-b_1}^{b_1} \int_{t(x_1)}^{0} \psi_1(x_1) R_1\psi_2(x_2, 0)\,\mathrm{d}x_2\mathrm{d}x_1.$$

Thus, we have

$$a^{\frac{1+\alpha}{2}} \int_{-b_1}^{b_1} \int_{t(x_1)}^{0} \psi_1(x_1)\left(\psi_2(0) + R_1\psi_2(x_2, 0)\right) \mathrm{d}x_2\mathrm{d}x_1$$

$$= -a^{\frac{1+\alpha}{2}} \left(\int_{-b_1}^{b_1} \psi_1(x_1)\psi_2(0)\,t(x_1)\right) \mathrm{d}x_1 + a^{\frac{1+\alpha}{2}} \int_{-b_1}^{b_1} \int_{t(x_1)}^{0} \psi_1(x_1) R_1\psi_2(x_2, 0)\,\mathrm{d}x_2\mathrm{d}x_1.$$

Concerning the first term, we get

$$\int_{-b_1}^{b_1} \psi_1(x_1)\psi_2(0)\left(g'(0) a^{1-\alpha} x_1 + \frac{1}{2} g''(\zeta) a^{2-\alpha} x_1^2\right) \mathrm{d}x_1$$

$$= a^{1-\alpha} \int_{-b_1}^{b_1} \psi_1(x_1)\psi_2(0) g'(0) x_1 \mathrm{d}x_1 + a^{C_\alpha} \int_{-b_1}^{b_1} \psi_1(x_1)\psi_2(0)\frac{1}{2} g''(\zeta) x_1^2 \mathrm{d}x_1$$

$$= a^{1-\alpha} \int_{-b_1}^{b_1} \psi_1(x_1)\psi_2(0) g'(0) x_1 \mathrm{d}x_1 + \mathcal{O}\left(a^{C_\alpha}\right).$$

with $C_\alpha > 1$. For the second term, we estimate $R_1\psi_2(x_2, 0)$. Since $\psi_2 \in C^2(\mathbb{R}) \cap L^2(\mathbb{R})$ is a bump function, we have that $\psi_2''(x_2) \leq M_1$ for all $x_2 \in (-t(x_1), t(x_1))$ and all $x_1 \in [-b_1, b_1]$ with some $M_1 < \infty$. According to [4], we can estimate the remainder $R_1\psi_2(x_2, 0)$ by

$$|R_1\psi_2(x_2, 0)| \leq \frac{1}{2} M_1 \left(t(x_1)\right)^2$$

$$= \frac{1}{2} M_1 \left(g'(0) a^{1-\alpha} x_1 + \frac{1}{2} g''(\zeta) a^{2-\alpha} x_1^2\right)^2$$

$$= \frac{1}{2} M_1 \left(g'(0)^2 a^{2(1-\alpha)} x_1^2 + g'(0) g''(\zeta) a^{(1-\alpha)(2-\alpha)} x_1^3 \right.$$

$$\left. + \frac{1}{4} g''(\zeta)^2 a^{2(2-\alpha)} x_1^4\right)$$

$$= \mathcal{O}\left(a^{2(1-\alpha)}\right)$$

for $a \to 0$. Together, this yields that (4.1) equals

$$-a^{\frac{1+\alpha}{2}} \left(a^{1-\alpha} \left(\int_{-b_1}^{b_1} \psi_1(x_1)\psi_2(0) g'(0) x_1 \mathrm{d}x_1\right) + \mathcal{O}\left(a^{2(1-\alpha)}\right)\right)$$

$$= -a^{\frac{3-\alpha}{2}} \left(\int_{-b_1}^{b_1} \psi_1(x_1)\psi_2(0) g'(0) x_1 \mathrm{d}x_1\right) + \mathcal{O}\left(a^{\frac{5-3\alpha}{2}}\right).$$

Analogously, we get for (4.2)

$$a^{\frac{3-\alpha}{2}} \left(\int_{-b_1}^{b_1} \psi_1(x_1)\psi_2(0) g'(0) x_1 \mathrm{d}x_1\right) + \mathcal{O}\left(a^{\frac{5-3\alpha}{2}}\right).$$

Subtracting (4.2) from (4.1) finally yields

$$\langle \chi_R, \psi_{a,0,0}\rangle = -a^{\frac{3-\alpha}{2}} \left(2 \int_{-b_1}^{b_1} \psi_1(x_1)\psi_2(0) g'(0) x_1 \mathrm{d}x_1\right) + \mathcal{O}\left(a^{\frac{5-3\alpha}{2}}\right).$$

Using the point-symmetry of $\psi_1(x_1)$ to $(0,0)$, we have that $\psi_1(x_1)x_1$ is axis-symmetric to $x_1 = 0$ and therefore $\int_{-b_1}^{0} \psi_1(x_1)x_1\mathrm{d}x_1 = \int_0^{b_1}\psi_1(x_1)x_1\mathrm{d}x_1$. With the assumption $\int_{-b_1}^{0}\psi_1(x_1)x_1\mathrm{d}x_1 \neq 0$ it follows that $\int_{-b_1}^{b_1}\psi_1(x_1)x_1\mathrm{d}x_1 \neq 0$. Thus, we finally get

$$\lim_{a\to 0+} a^{-\frac{3-\alpha}{2}}|\langle \chi_R, \psi_{a,0,0}\rangle| > 0.$$

ii. Again, we analyze the situation for $\langle \chi_R, \psi_{a,0,0}\rangle$. As before, we apply the transformation theorem to get

$$\langle \chi_R, \psi_{a,0,0}\rangle = a^{\frac{1+\alpha}{2}}\int_{A_{a,\alpha}^{-1}R}\psi(x)\,\mathrm{d}x.$$

As shown by Kutyniok and Petersen [78], the boundary curve of R is given by $\vec{\alpha} = (\vec{\alpha}_1, \vec{\alpha}_2)^T$ and can be expressed as a function $g := \vec{\alpha}_1 \circ \vec{\alpha}_2^{-1}$. Similarly as in [78] but adjusted to the scaling matrix $A_{a,\alpha}$ instead of A_a, we have

$$A_{a,\alpha}^{-1}R \cap \operatorname{supp}\psi = \{(x_1, x_2) \in \operatorname{supp}\psi \colon ax_1 \leq g(a^\alpha x_2)\}.$$

A Taylor expansion of g at 0 with utilization of the Lagrange form of the remainder provides

$$
\begin{aligned}
\langle \chi_R, \psi_{a,0,0}\rangle &= a^{\frac{1+\alpha}{2}}\int_{ax_1 \leq g(0)+g'(0)a^\alpha x_2 + R_1 g(a^\alpha x_2, 0)}\psi(x)\,\mathrm{d}x \\
&= a^{\frac{1+\alpha}{2}}\int_{ax_1 \leq g'(0)a^\alpha x_2 + \frac{1}{2}g''(\zeta)a^{2\alpha}x_2^2}\psi(x)\,\mathrm{d}x \\
&= a^{\frac{1+\alpha}{2}}\int_{x_1 \leq g'(0)a^{\alpha-1}x_2 + \frac{1}{2}g''(\zeta)a^{2\alpha-1}x_2^2}\psi(x)\,\mathrm{d}x
\end{aligned}
$$

with $0 < \zeta < a^\alpha x_2$. With $t(x_2) := g'(0)a^{\alpha-1}x_2 + \frac{1}{2}g''(\zeta)a^{2\alpha-1}x_2^2$, we get

$$
\begin{aligned}
t(x_2) &= \frac{1}{n_1}\vec{\alpha}_1'(t_0)a^{\alpha-1}x_2 + \frac{1}{2}g''(\zeta)a^{2\alpha-1}x_2^2 \\
&= \frac{n_2}{n_1}a^{\alpha-1}x_2 + \frac{1}{2}g''(\zeta)a^{2\alpha-1}x_2^2,
\end{aligned}
$$

since $\left(\alpha_2^{-1}\right)'(0) = 1/n_1$ and $\vec{\alpha}_1'(t_0) = n_2$. Using the separability of local precision shearlets, we have

$$\langle \chi_R, \psi_{a,0,0}\rangle = a^{\frac{1+\alpha}{2}}\int_{x_1 \leq t(x_2)}\psi_1(x_1)\psi_2(x_2)\,\mathrm{d}x.$$

We split the integral

$$
\begin{aligned}
\int_{x_1 \leq t(x_2)}\psi_1(x_1)\psi_2(x_2)\,\mathrm{d}x &= \int_{\substack{t(x_2)\geq 0 \\ x_1 \leq t(x_2)}}\psi_1(x_1)\psi_2(x_2)\,\mathrm{d}x + \int_{\substack{t(x_2)\leq 0 \\ x_1 \leq t(x_2)}}\psi_1(x_1)\psi_2(x_2)\,\mathrm{d}x \\
&= \int_{\substack{t(x_2)\geq 0 \\ x_1 \leq 0}}\psi_1(x_1)\psi_2(x_2)\,\mathrm{d}x + \int_{0\leq x_1 \leq t(x_2)}\psi_1(x_1)\psi_2(x_2)\,\mathrm{d}x \\
&\quad + \int_{x_1 \leq t(x_2)\leq 0}\psi_1(x_1)\psi_2(x_2)\,\mathrm{d}x \\
&= \int_{x_1 \leq 0}\psi_1(x_1)\psi_2(x_2)\,\mathrm{d}x \\
&\quad + \int_{0\leq x_1 \leq t(x_2)}\psi_1(x_1)\psi_2(x_2)\,\mathrm{d}x \qquad (4.3)
\end{aligned}
$$

$$-\int_{t(x_2)\leq x_1\leq 0}\psi_1\left(x_1\right)\psi_2\left(x_2\right)\mathrm{d}x. \tag{4.4}$$

Considering (4.3), we apply a Taylor expansion of ψ_1 at 0 to get

$$
\begin{aligned}
\int_{0\leq x_1\leq t(x_2)}\psi_1\left(x_1\right)\psi_2\left(x_2\right)\mathrm{d}x &= \int_{-b_2}^{b_2}\psi_2\left(x_2\right)\int_0^{t(x_2)}\psi_1\left(0\right)+\psi_1'\left(0\right)x_1\\
&\quad +R_1\psi_1\left(x_1,0\right)\mathrm{d}x_1\mathrm{d}x_2\\
&= \int_{-b_2}^{b_2}\psi_2\left(x_2\right)\int_0^{t(x_2)}\psi_1'\left(0\right)x_1+R_1\psi_1\left(x_1,0\right)\mathrm{d}x_1\mathrm{d}x_2\\
&= \int_{-b_2}^{b_2}\psi_2\left(x_2\right)\frac{1}{2}\psi_1'\left(0\right)\left(t\left(x_2\right)\right)^2\\
&\quad +\int_0^{t(x_2)}\psi_2\left(x_2\right)R_1\psi_1\left(x_1,0\right)\mathrm{d}x_1\mathrm{d}x_2
\end{aligned}
$$

Similar as in i., we get

$$
\begin{aligned}
\left|R_1\psi_1\left(x_1,0\right)\right| &\leq \frac{1}{2}M_1\left(t\left(x_2\right)\right)^2\\
&= \frac{1}{2}M_1\left(\frac{n_2}{n_1}a^{\alpha-1}x_2+\frac{1}{2}g''\left(\zeta\right)a^{2\alpha-1}x_2^2\right)^2\\
&\leq \frac{1}{2}M_1\left(a^\alpha x_2+\frac{1}{2}g''\left(\zeta\right)a^{2\alpha-1}x_2^2\right)^2\\
&= \mathcal{O}\left(a^{2\alpha}\right)
\end{aligned}
$$

with $M_1<\infty$, since in our situation we have $n_2/n_1\leq a$. Therefore, we have

$$\int_{0\leq x_1\leq t(x_2)}\psi_1\left(x_1\right)\psi_2\left(x_2\right)\mathrm{d}x=\mathcal{O}\left(a^{2\alpha}\right).$$

The term $\int_{t(x_2)\leq x_1\leq 0}\psi_1\left(x_1\right)\psi_2\left(x_2\right)\mathrm{d}x$ is approximated analogously such that we get

$$a^{\frac{1+\alpha}{2}}\int_{x_1\leq t(x_2)}\psi_1\left(x_1\right)\psi_2\left(x_2\right)\mathrm{d}x=a^{\frac{1+\alpha}{2}}\int_{x_1\leq 0}\psi_1\left(x_1\right)\psi_2\left(x_2\right)\mathrm{d}x+\mathcal{O}\left(a^{\frac{5\alpha+1}{2}}\right).$$

Due to the symmetry properties of a local precision shearlet, we have $\int_{x_1\leq 0}\psi_1\left(x_1\right)\psi_2\left(x_2\right)\mathrm{d}x\neq 0$ and therefore

$$\lim_{a\to 0+}a^{-\frac{1+\alpha}{2}}\left|\langle\chi_R,\psi_{a,0,0}\rangle\right|>0.$$

Since ψ is a bounded compactly supported shearlet we can use Theorem 2.21 to state an explicit limit value for $\mathcal{SH}_\psi\chi_R\left(a,s,p\right)$ in case of $\alpha=1/2$. From Theorem 2.21 we know that for $\delta>0$, $\|p-p_i\|>0$ for all corner points p_i of ∂R there exists a constant C_δ such that for all $a\in(0,1]$ we have

$$a^{\frac{3}{4}}\int_S\psi\left(x\right)\mathrm{d}x-C_\delta a^{\frac{5}{4}}\leq\mathcal{SH}_\psi\chi_R\left(a,s,p\right)\leq a^{\frac{3}{4}}\int_S\psi\left(x\right)\mathrm{d}x+C_\delta a^{\frac{5}{4}}$$

if $s\in B_a\left(s_0\right)=\left\{s\in\mathbb{R}\colon |s-s_0|\leq a\right\}$. Therefore we have

$$\lim_{a\to 0+}\mathcal{SH}_\psi\chi_R\left(a,s,p\right)=a^{\frac{3}{4}}\int_S\psi\left(x\right)\mathrm{d}x+\mathcal{O}\left(a^{\frac{5}{4}}\right)$$

and consequently

$$\lim_{a\to 0+}a^{-\frac{3}{4}}\mathcal{SH}_\psi\chi_R\left(a,s,p\right)=\int_S\psi\left(x\right)\mathrm{d}x.$$

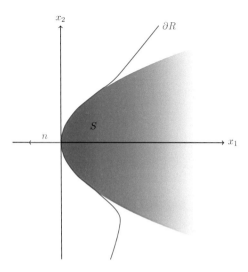

Figure 4.1: Illustration of a regular point of ∂R at $(0,0)$ and the corresponding integral area S.

$\square$

Remark 4.3.

  i. For the statement above, we need a local precision shearlet to fulfill the condition $\int_{-b_1}^{0} \psi_1(x_1) x_1 \mathrm{d}x_1 \neq 0$. Also in the subsequent statements about edge point characterization using local precision shearlets, we will use this condition. We will discuss its consequential limitation on suitable shearlets in Section 4.2. However, we will show that this limitation is in harmony with the observed behavior in the practical application of an edge detection algorithm.

  ii. If s corresponds to the normal direction, an explicit limit value of $\mathcal{SH}_{\psi}\chi_R(a,s,p)$ can be only derived in case of parabolic scaling, i.e. $\alpha = 1/2$. Only in this case a Taylor expansion of $1/a\vec{\alpha}(\vec{\alpha}_2^{-1}(a^{\alpha}x_2))$ provides the finding of the integration area S. Here, the influence of a vanishes in the second order component of the Taylor expansion. For more details, see [78].

We now turn to the characterization of corner points $p \in \partial R$ of the first type. When s does not correspond to the normal direction at p, we derive that $\lim_{a\to 0+} a^{-\frac{3-\alpha}{4}} |\mathcal{SH}_{\psi}\chi_R(a,s,p)| > 0$. For $\alpha = 1/2$, this matches the result of Theorem 2.21. Although the properties of a local precision shearlet differ from the conditions of the shearlet in Theorem 2.21, i.e. $\psi_2(0) = 0$ and $\psi_2'(0) \neq 0$. In case s corresponds to the normal direction, we find again the decay rate of $a^{\frac{1+\alpha}{2}}$ for $a \to 0$. Again, this matches the decay rate in Theorem 2.21 for $\alpha = 1/2$. In this case, we even get a stronger result than in Theorem 2.21 by the statement of an explicit limit value of $a^{-\frac{3}{4}}\mathcal{SH}_{\psi}\chi_R(a,s,p)$ for $a \to 0+$. In contrast, Theorem 2.21 provides two possible limit values.

Theorem 4.4. *Let* $\psi \in L^2(\mathbb{R}^2)$ *be a local precision shearlet with* $\operatorname{supp} \psi \subseteq [-b_1, b_1] \times [-b_2, b_2] \subset \mathbb{R}^2$, $\int_{-b_1}^0 \psi_1(x_1) x_1 \mathrm{d}x_1 \neq 0$ *and* $\psi_2 \in C^2(\mathbb{R}) \cap L^2(\mathbb{R})$ *as well as* $R \subset \mathbb{R}^2$, *with boundary* ∂R *of length* L, *be smooth except for finitely many corner points. Let furthermore* $\vec{\alpha}(t_0) = p \in \partial R$, $t_0 \in (0, L)$, *be a corner point of first type.*

i. If s does not correspond to a normal direction of ∂R or a tangent direction, then we have

$$\lim_{a \to 0+} a^{-\frac{3-\alpha}{2}} |\mathcal{SH}_\psi \chi_R(a, s, p)| > 0.$$

ii. If s_0 corresponds to a normal direction of ∂R, then we have

$$\lim_{a \to 0+} a^{-\frac{1+\alpha}{2}} |\mathcal{SH}_\psi \chi_R(a, s, p)| > 0.$$

If furthermore $\alpha = 1/2$, then

$$\lim_{a \to 0+} a^{-\frac{3}{4}} \mathcal{SH}_\psi \chi_R(a, s, p) = \int_{\widetilde{S}} \psi(x) \, \mathrm{d}x \quad \textit{if } s \in B_a(s_0),$$

with $B_a(s_0) = \{s \in \mathbb{R}: |s - s_0| \leq a\}$,

$$\widetilde{S} := \left\{ (x_1, x_2) \in \operatorname{supp} \psi \colon x_1 \leq \frac{\vec{\alpha}_1''(t_0) - s\vec{\alpha}_2''(t_0)}{2\rho(s)^2} x_2^2, x_2 \geq 0 \right\}$$

and $\rho(s) := \cos(\operatorname{atan}(s))$.

Proof. We show the two statements separately.

i. According to the proof of Theorem 2.21 [78], we have

$$
\begin{aligned}
\langle \chi_R, \psi_{a,0,0} \rangle &= a^{\frac{3}{4}} \left\langle \chi_{A_a^{-1}\widetilde{T}}, \psi_{a,0,0} \right\rangle + a^{\frac{3}{4}} \left\| \chi_{A_a^{-1}T} - \chi_{A_a^{-1}\widetilde{T}} \right\|_1 \\
&= a^{\frac{3}{4}} \left\langle \chi_{A_a^{-1}\widetilde{T}}, \psi_{a,0,0} \right\rangle + a^{\frac{3}{4}} \mathcal{O}\left(a^{\frac{3}{2}}\right) \\
&= a^{\frac{3}{4}} \left\langle \chi_{A_a^{-1}\widetilde{T}}, \psi_{a,0,0} \right\rangle + O\left(a^{\frac{9}{4}}\right)
\end{aligned}
$$

with

$$
\begin{aligned}
T &:= \left\{ (x_1, x_2) \in \operatorname{supp} \psi \colon x_1 \leq 0, \ g^-(x_1) \leq x_2 \leq g^+(x_1) \right\}, \\
\widetilde{T} &:= \left\{ (x_1, x_2) \in \operatorname{supp} \psi \colon x_1 \leq 0, \ (g^-)'(0) x_1 \leq x_2 \leq \left(g^+\right)'(0) x_1 \right\}, \\
g^+ &:= \vec{\alpha}_{2|t \geq t_0} \circ \vec{\alpha}_1^{-1}, \\
g^- &:= \vec{\alpha}_{2|t \leq t_0} \circ \vec{\alpha}_1^{-1},
\end{aligned}
$$

for the case of $\alpha = 1/2$. The case $(g^+)'(0) \leq 0 < (g^-)'(0)$ is considered but according to [78], the same end result is achieved for different constellations of g^+ and g^-. For a general $\alpha \in [1/2, 1)$, we analogously have

$$\langle \chi_R, \psi_{a,0,0} \rangle = a^{\frac{1+\alpha}{2}} \left\langle \chi_{A_{a,\alpha}^{-1}\widetilde{T}}, \psi \right\rangle + a^{\frac{1+\alpha}{2}} \left\| \chi_{A_{a,\alpha}^{-1}T} - \chi_{A_{a,\alpha}^{-1}\widetilde{T}} \right\|_{L^1(\mathbb{R}^2)}.$$

With the specifications of T and $\widetilde{T}$ we get

$$\left\| \chi_{A_{a,\alpha}^{-1}T} - \chi_{A_{a,\alpha}^{-1}\widetilde{T}} \right\|_{L^1(\mathbb{R}^2)} = \int_{A_{a,\alpha}^{-1}T} 1 \mathrm{d}x - \int_{A_{a,\alpha}^{-1}\widetilde{T}} 1 \mathrm{d}x$$

$$= \int_{-b_1}^{0} \int_{g^-(x_1)a^{1-\alpha}x_1}^{g^+(x_1)a^{1-\alpha}x_1} 1 \, \mathrm{d}x_2 \mathrm{d}x_1 - \int_{-b_1}^{0} \int_{(g^-)'(0)a^{1-\alpha}x_1}^{(g^+)'(0)a^{1-\alpha}x_1} 1 \, \mathrm{d}x_2 \mathrm{d}x_1.$$

By a Taylor expansion of g^- and g^+ at 0 with Lagrange form of the remainder we have $g^+(x_1) = (g^+)'(0) a^{1-\alpha} x_1 + \frac{1}{2} (g^+)'(\zeta) a^{2-\alpha} x_1$ and $g^-(x_1) = (g^-)'(0) a^{1-\alpha} x_1 + \frac{1}{2} (g^-)'(\tilde{\zeta}) a^{2-\alpha} x_1$ with ζ and $\tilde{\zeta}$ between 0 and x_1. Thus, we get

$$\left\| \chi_{A_{a,\alpha}T} - \chi_{A_{a,\alpha}\tilde{T}} \right\|_{L^1(\mathbb{R}^2)} = \int_{-b_1}^{0} \int_{(g^-)'(0)a^{1-\alpha}x_1 + \frac{1}{2}(g^-)'(\tilde{\zeta})a^{2-\alpha}x_1}^{(g^+)'(0)a^{1-\alpha}x_1 + \frac{1}{2}(g^+)'(\zeta)a^{2-\alpha}x_1} 1 \, \mathrm{d}x_2 \mathrm{d}x_1$$

$$- \int_{-b_1}^{0} \int_{(g^-)'(0)a^{1-\alpha}x_1}^{(g^+)'(0)a^{1-\alpha}x_1} 1 \, \mathrm{d}x_2 \mathrm{d}x_1.$$

$$= \frac{1}{2} a^{2-\alpha} \int_{-b_1}^{0} \left((g^+)'(\zeta) - (g^-)'(\tilde{\zeta}) \right) x_1 \mathrm{d}x_1$$

$$= \mathcal{O}\left(a^{2-\alpha} \right)$$

for for $a \to 0$. Thus, we have

$$\langle \chi_R, \psi_{a,0,0} \rangle = a^{\frac{1+\alpha}{2}} \left\langle \chi_{A_{a,\alpha}^{-1}\tilde{T}}, \psi \right\rangle + a^{\frac{1+\alpha}{2}} \mathcal{O}\left(a^{2-\alpha} \right)$$

$$= a^{\frac{1+\alpha}{2}} \left\langle \chi_{A_{a,\alpha}^{-1}\tilde{T}}, \psi \right\rangle + \mathcal{O}\left(a^{\frac{5-\alpha}{2}} \right)$$

$$= a^{\frac{1+\alpha}{2}} \int_{A_{a,\alpha}^{-1}\tilde{T}} \psi(x) \, \mathrm{d}x + \mathcal{O}\left(a^{\frac{5-\alpha}{2}} \right).$$

Now, we make use of the specific properties of a local precision shearlet. Due to its separability and with a Taylor expansion of ψ_2 at 0 we get

$$\langle \chi_R, \psi_{a,0,0} \rangle = a^{\frac{1+\alpha}{2}} \int_{-b_1}^{0} \int_{(g^-)'(0)a^{1-\alpha}x_1}^{(g^+)'(0)a^{1-\alpha}x_1} \psi_1(x_1) \psi_2(x_2) \, \mathrm{d}x_2 \mathrm{d}x_1 + \mathcal{O}\left(a^{\frac{5-\alpha}{2}} \right)$$

$$= a^{\frac{1+\alpha}{2}} \int_{-b_1}^{0} \int_{(g^-)'(0)a^{1-\alpha}x_1}^{(g^+)'(0)a^{1-\alpha}x_1} \psi_1(x_1) \psi_2(0) + \psi_2'(0) x_2$$

$$+ R_1 \psi_2(x_2, 0) \, \mathrm{d}x_2 \mathrm{d}x_1 + \mathcal{O}\left(a^{\frac{5-\alpha}{2}} \right)$$

for $a \to 0$. Since ψ_2 is a bump function axis-symmetric to $x_2 = 0$ we have $\psi_2'(0) = 0$. Thus, we have

$$\langle \chi_R, \psi_{a,0,0} \rangle = a^{\frac{1+\alpha}{2}} \int_{-b_1}^{0} \int_{(g^-)'(0)a^{1-\alpha}x_1}^{(g^+)'(0)a^{1-\alpha}x_1} \psi_1(x_1) \psi_2(0) \, \mathrm{d}x_2 \mathrm{d}x_1$$

$$+ a^{\frac{1+\alpha}{2}} \int_{-b_1}^{0} \int_{(g^-)'(0)a^{1-\alpha}x_1}^{(g^+)'(0)a^{1-\alpha}x_1} \psi_1(x_1) R_1 \psi_2(x_2, 0) \, \mathrm{d}x_2 \mathrm{d}x_1 + \mathcal{O}\left(a^{\frac{5-\alpha}{2}} \right).$$

Concerning the first term, integrating along x_2 yields

$$a^{\frac{1+\alpha}{2}} \int_{-b_1}^{0} \int_{(g^-)'(0)a^{1-\alpha}x_1}^{(g^+)'(0)a^{1-\alpha}x_1} \psi_1(x_1) \psi_2(0) \, \mathrm{d}x_2 \mathrm{d}x_1$$

$$= a^{\frac{3-\alpha}{2}} \left(\int_{-b_1}^{0} \psi_1(x_1) \psi_2(0) \left((g^+)'(0) - (g^-)'(0) \right) x_1 \mathrm{d}x_1 \right).$$

For the second term, we estimate $R_1 \psi_2(x_2, 0)$. Since $\psi_2 \in C^2(\mathbb{R}) \cap L^2(\mathbb{R})$ is a bump function, we have that $\psi_2''(x_2) \le M_1$ for all $x_2 \in (-r(x_1), r(x_1))$ and all $x_1 \in [-b_1, 0]$ with

$r(x_1) = a^{1-\alpha} x_1 \max\left(\left|(g^-)'(0)\right|, \left|(g^+)'(0)\right|\right)$ and some $M_1 < \infty$. From [4], we estimate the remainder $R_1 \psi_2(x_2, 0)$ by

$$
\begin{aligned}
|R_1 \psi_2(x_2, 0)| &\leq \frac{1}{2} M_1 \left(r(x_1)\right)^2 \\
&= \frac{1}{2} M_1 \left(a^{1-\alpha} x_1 \max\left(\left|(g^-)'(0)\right|, \left|(g^+)'(0)\right|\right)\right)^2 \\
&= \frac{1}{2} M_1 \left(a^{2(1-\alpha)} x_1^2 \max\left(\left|(g^-)'(0)\right|, \left|(g^+)'(0)\right|\right)^2\right) \\
&= \mathcal{O}\left(a^{2(1-\alpha)}\right)
\end{aligned}
$$

for $a \to 0$. Putting the pieces together, we get

$$
\begin{aligned}
\langle \chi_R, \psi_{a,0,0} \rangle &= a^{\frac{3-\alpha}{2}} \int_{-b_1}^{0} \psi_1(x_1) \psi_2(0) \left(\left(g^+\right)'(0) - \left(g^-\right)'(0)\right) x_1 \mathrm{d}x_1 \\
&\quad + a^{\frac{1+\alpha}{2}} \mathcal{O}\left(a^{2(1-\alpha)}\right) + \mathcal{O}\left(a^{\frac{5-\alpha}{2}}\right) \\
&= a^{\frac{3-\alpha}{2}} \int_{-b_1}^{0} \psi_1(x_1) \psi_2(0) \left(\left(g^+\right)'(0) - \left(g^-\right)'(0)\right) x_1 \mathrm{d}x_1 + \mathcal{O}\left(a^{\frac{5-3\alpha}{2}}\right).
\end{aligned}
$$

With the assumption $\int_{-b_1}^{0} \psi_1(x_1) x_1 \mathrm{d}x_1 \neq 0$, we finally have

$$
\lim_{a \to 0+} a^{-\frac{3-\alpha}{2}} |\langle \chi_R, \psi_{a,0,0} \rangle| > 0.
$$

ii. As described by [78], we can write $\chi_R = \chi_{R_1} + \chi_{R_2}$ where ∂R_1 as well as ∂R_2 have a corner point of the first type at p. Furthermore, the normals of ∂R_1 are perpendicular, where one corresponds to s. Finally, none of the normals of ∂R_2 corresponds to s. From i. we know $|\langle \chi_{R_2}, \psi_{a,0,0} \rangle| = \mathcal{O}(a^{\frac{3-\alpha}{2}})$ for $a \to 0$ and therefore

$$
\begin{aligned}
|\langle \chi_R, \psi_{a,0,0} \rangle| &= |\langle \chi_{R_1}, \psi_{a,0,0} \rangle + \langle \chi_{R_2}, \psi_{a,0,0} \rangle| \\
&= |\langle \chi_{R_1}, \psi_{a,0,0} \rangle| + \mathcal{O}\left(a^{\frac{3-\alpha}{2}}\right)
\end{aligned}
$$

With the same method as in the proof of 4.2 ii., we derive

$$
\lim_{a \to 0+} a^{-\frac{1+\alpha}{2}} |\mathcal{SH}_\psi \chi_{R_1}(a, s, p)| > 0.
$$

For the special case $\alpha = 1/2$, we know from Theorem 2.21 [78] that for compactly supported shearlets we have

$$
\lim_{a \to 0+} a^{-\frac{3}{4}} \langle \chi_R, \psi_{a,s,p} \rangle \in \left\{ \int_{\widetilde{S}^+} \psi(x) \, \mathrm{d}x, \int_{\widetilde{S}^-} \psi(x) \, \mathrm{d}x \right\} \quad \text{if } s \in B_a(s_0),
$$

with $\widetilde{S}^+ := S \cap \{x \in \mathbb{R}^2 : x_2 \geq 0\}$ and $\widetilde{S}^- := S \cap \{x \in \mathbb{R}^2 : x_2 < 0\}$. Since ψ is separable and ψ_2 is axis-symmetric to $x_2 = 0$, we get

$$
\begin{aligned}
\int_{\widetilde{S}^+} \psi(x) \, \mathrm{d}x &= \int\int_{\left\{ x \in \mathrm{supp}\, \psi : \, x_1 \leq \frac{(\tilde{\alpha}_1''(t_0) - s\tilde{\alpha}_2''(t_0))}{2\rho(s)^2} x_2^2, x_2 \geq 0 \right\}} \psi_1(x_1) \psi_2(x_2) \, \mathrm{d}x \\
&= \int\int_{\left\{ x \in \mathrm{supp}\, \psi : \, x_1 \leq \frac{(\tilde{\alpha}_1''(t_0) - s\tilde{\alpha}_2''(t_0))}{2\rho(s)^2} (-x_2)^2, x_2 \leq 0 \right\}} \psi_1(x_1) \psi_2(-x_2) \, \mathrm{d}x
\end{aligned}
$$

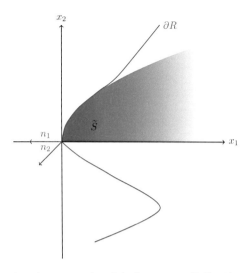

Figure 4.2: Illustration of a corner point of the first type at $(0,0)$ with one normal $\vec{n}_1 = (-1,0)$ and the corresponding integral area $\widetilde{S}$.

$$= \int_{\widetilde{S}-} \psi_1 (x_1) \psi_2 (-x_2) \, \mathrm{d}x.$$

Due to the axis-symmetry of ψ_2, i.e. $\psi_2 (x_2) = \psi_2 (-x_2)$, we have

$$\int_{\widetilde{S}+} \psi (x) \, \mathrm{d}x = \int_{\widetilde{S}-} \psi (x) \, \mathrm{d}x$$

and therefore

$$\lim_{a \to 0+} a^{-\frac{3}{4}} \langle \chi_R, \psi_{a,s,p} \rangle = \int_{\widetilde{S}} \psi (x) \, \mathrm{d}x.$$

$\square$

Finally, we examine the situation for corner points of the second type. If s does not correspond to the normal direction of ∂R at p, we obtain the decay rate of $a^{\frac{3-\alpha}{2}}$ for $a \to 0$. For $\alpha = 1/2$, this is a different result than in Theorem 2.21 due to differing properties of a local precision shearlet in comparison to the shearlet described in Theorem 2.21. However, our result is suitable to the other results given in this section. If s corresponds to the normal direction, the properties of a local precision shearlet enable us to obtain an explicit limit value. For this case, Theorem 2.21 does not provide any result.

Theorem 4.5. *Let* $\psi \in L^2 (\mathbb{R}^2)$ *be a local precision shearlet with* $\mathrm{supp}\, \psi \subseteq [-b_1, b_1] \times [-b_2, b_2] \subset \mathbb{R}$ *and* $\int_{-b_1}^{0} \psi_1 (x_1) x_1 \mathrm{d}x_1 \neq 0$ *as well as* $R \subset \mathbb{R}^2$, *with boundary* ∂R *of length* L, *be smooth except for finitely many corner points. Let furthermore* $\vec{\alpha} (t_0) = p \in \partial R$, $t_0 \in (0, L)$, *be a corner point of second type.*

i. If s does not correspond to a normal direction of ∂R or a tangent direction, then we have

$$\lim_{a \to 0+} a^{-\frac{3-\alpha}{2}} |\mathcal{SH}_\psi \chi_R (a, s, p)| > 0.$$

ii. *If s_0 corresponds to the normal direction $\vec{n}(t_0)$ of ∂R, then we have*

$$\lim_{a \to 0+} a^{-\frac{1+\alpha}{2}} |\mathcal{SH}_\psi \chi_R(a,s,p)| > 0.$$

If furthermore $\alpha = 1/2$, then

$$\lim_{a \to 0+} a^{-\frac{3}{4}} \mathcal{SH}_\psi \chi_R(a,s,p) = \int_{\widetilde{S}^- \cup \widetilde{S}^+} \psi(x)\,\mathrm{d}x \quad \text{if } s \in B_a(s_0),$$

with $B_a(s_0) = \{s \in \mathbb{R} \colon |s - s_0| \leq a\}$,

$$\widetilde{S}^- := \left\{ (x_1, x_2) \in \operatorname{supp}\psi \colon x_1 \leq \frac{\vec{\alpha}_1''\left(t_0^-\right) - s\vec{\alpha}_2''\left(t_0^-\right)}{2\rho(s)^2} x_2^2, x_2 < 0 \right\},$$

$$\widetilde{S}^+ := \left\{ (x_1, x_2) \in \operatorname{supp}\psi \colon x_1 \leq \frac{\vec{\alpha}_1''\left(t_0^+\right) - s\vec{\alpha}_2''\left(t_0^+\right)}{2\rho(s)^2} x_2^2, x_2 \geq 0 \right\}$$

and $\rho(s) := \cos(\operatorname{atan}(s))$.

Proof. The two statements are shown as follows:

i. Again, we first follow the procedure of [78] and examine the behavior for the special case $s = 0$ and $p = 0$ while the result can directly be transferred to the general case. According to the proof of Theorem 2.21 [78], $\vec{\alpha}$ is locally given as the graph of a function $g \colon [-\varepsilon, \varepsilon] \to [-\varepsilon, \varepsilon]$ such that for a small enough

$$\begin{aligned}
\langle \chi_R, \psi_{a,0,0} \rangle &= \int_{x_2 \geq g(x_1)} \psi_{a,0,0}\,\mathrm{d}x \\
&= \int_{0 \geq x_2 \geq g'(0)x_1 + \frac{1}{2}(g^-)''(0)x_1^2} \psi_{a,0,0}\,\mathrm{d}x \\
&\quad + \int_{x_2 \geq g'(0)x_1 + \frac{1}{2}(g^+)''(0)x_1^2 \geq 0} \psi_{a,0,0}\,\mathrm{d}x + \mathcal{O}\left(a^3\right),
\end{aligned}$$

where the part of g for $x_1 \leq 0$ is denoted by g^- and the part for $x_1 > 0$ by g^+. Since p is a corner point of the second type, we have $(g^-)''(0) \neq \pm (g^+)''(0)$. With an application of the transformation theorem and by setting $t^-(x_1) := g'(0) a^{1-\alpha} x_1 + \frac{1}{2} (g^-)''(0) a^{2-\alpha} x_1^2$, $t^+(x_1) := g'(0) a^{1-\alpha} x_1 + \frac{1}{2} (g^+)''(0) a^{2-\alpha} x_1^2$ we have

$$\begin{aligned}
\langle \chi_R, \psi_{a,0,0} \rangle &= a^{\frac{1+\alpha}{2}} \int_{0 \geq x_2 \geq t^-(x_1)} \psi_1(x_1) \psi_2(x_2)\,\mathrm{d}x \\
&\quad + a^{\frac{1+\alpha}{2}} \int_{x_2 \geq t^+(x_1) \geq 0} \psi_1(x_1) \psi_2(x_2)\,\mathrm{d}x + \mathcal{O}\left(a^3\right).
\end{aligned}$$

Applying the same method as in the proof of Theorem 4.2 i., we get

$$\lim_{a \to 0+} a^{-\frac{3-\alpha}{2}} |\langle \chi_R, \psi_{a,0,0} \rangle| > 0.$$

ii. We split $\chi_R = \chi_{R_1} + \chi_{R_2}$ such that ∂R_1 and ∂R_2 both have a corner point of the first type at p with each one normal direction that corresponds to s_0 and the other perpendicular to it. Without loss of generality, we assume that $\vec{\alpha}(t_0^-) \in \partial R_1$ and $\vec{\alpha}(t_0^+) \in \partial R_2$. In the opposite case, we would just switch the indices to get to the same result. From Theorem 4.4

we know $\lim_{a\to 0+} a^{-\frac{1+\alpha}{2}} |\mathcal{SH}_\psi \chi_{R_1}(a,s,p)| > 0$, $\lim_{a\to 0+} a^{-\frac{1+\alpha}{2}} |\mathcal{SH}_\psi \chi_{R_2}(a,s,p)| > 0$ and therefore

$$\lim_{a\to 0+} a^{-\frac{1+\alpha}{2}} |\mathcal{SH}_\psi \chi_R(a,s,p)| > 0.$$

For the special case of $\alpha = 1/2$, Theorem 4.4 furthermore provides

$$\lim_{a\to 0+} a^{-\frac{3}{4}} \langle \chi_{R_1}, \psi_{a,s,p} \rangle = \int_{\widetilde{S}^-} \psi(x)\,\mathrm{d}x$$

and

$$\lim_{a\to 0+} a^{-\frac{3}{4}} \langle \chi_{R_2}, \psi_{a,s,p} \rangle = \int_{\widetilde{S}^+} \psi(x)\,\mathrm{d}x$$

with $s \in B_a(s_0)$,

$$\widetilde{S}^- := \left\{ (x_1,x_2) \in \operatorname{supp}\psi \colon x_1 \leq \frac{\vec{\alpha}_1''\left(t_0^-\right) - s\vec{\alpha}_2''\left(t_0^-\right)}{2\rho(s)^2} x_2^2, x_2 \geq 0 \right\}$$

and

$$\widetilde{S}^+ := \left\{ (x_1,x_2) \in \operatorname{supp}\psi \colon x_1 \leq \frac{\vec{\alpha}_1''\left(t_0^-\right) - s\vec{\alpha}_2''\left(t_0^-\right)}{2\rho(s)^2} x_2^2, x_2 \geq 0 \right\}.$$

Therefore, we have

$$\lim_{a\to 0+} a^{-\frac{3}{4}} \langle \chi_R, \psi_{a,s,p} \rangle = \lim_{a\to 0+} a^{-\frac{3}{4}} \langle \chi_{R_1}, \psi_{a,s,p} \rangle + \lim_{a\to 0+} a^{-\frac{3}{4}} \langle \chi_{R_2}, \psi_{a,s,p} \rangle$$
$$= \int_{\widetilde{S}^-} \psi(x)\,\mathrm{d}x + \int_{\widetilde{S}^+} \psi(x)\,\mathrm{d}x.$$

As can be seen in the proof of Theorem 4.4, the axis-symmetry of ψ_2 enables us to reformulate $\widetilde{S}^-$ by

$$\widetilde{S}^- := \left\{ (x_1,x_2) \in \operatorname{supp}\psi \colon x_1 \leq \frac{1}{2\rho(s)^2} \left(\vec{\alpha}_1''\left(t_0^-\right) - s\vec{\alpha}_2''\left(t_0^-\right) \right) x_2^2, x_2 < 0 \right\}.$$

Therefore, we finally have

$$\lim_{a\to 0+} a^{-\frac{3}{4}} \langle \chi_R, \psi_{a,s,p} \rangle = \int_{\widetilde{S}^-\cup\widetilde{S}^+} \psi(x)\,\mathrm{d}x.$$

$\square$

Summarizing, if one uses a local precision shearlet ψ for edge detection, $\mathcal{SH}_\psi \chi_R(a,s,p)$ will decay for $a \to 0$ with $\mathcal{O}(a^{\frac{3-\alpha}{2}})$ if the shear parameter s does not correspond to the normal direction at $p \in \partial R$. The condition we need a local precision shearlet to fulfill is

$$\int_{-b_1}^{0} \psi_1(x_1)\, x_1 \mathrm{d}x_1 \neq 0 \tag{4.5}$$

with $\operatorname{supp}\psi \subseteq [-b_1, b_1] \times [-b_2, b_2]$. This holds true for regular as well as corner points. If s corresponds to the normal direction $\mathcal{SH}_\psi \chi_R(a,s,p)$ will decay with $\mathcal{O}(a^{\frac{1+\alpha}{2}})$. For $\alpha = 1/2$, we can distinguish regular points, corner points of the first and the second type by the limit value which can be computed specifically. As a second benefit in case of $\alpha = 1/2$, the difference in the decay rates $a^{\frac{3-\alpha}{2}}$ and $a^{\frac{1+\alpha}{2}}$ is more significant than for higher values of α. Although we have these advantages for $\alpha = 1/2$, higher values of α enable us to use only few shearlets for fine scales such that we still cover the whole frequency plane. We will see in 5.5.2 that the consideration of $\alpha > 1/2$ leads to improved pedestrian detection rates.

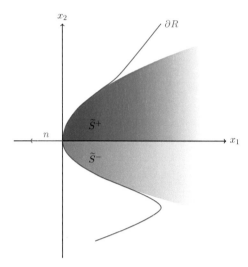

Figure 4.3: Illustration of a corner point of the second type at $(0,0)$ and the corresponding integral area $\widetilde{S}^- \cup \widetilde{S}^+$.

4.2   Suitable Shearlets

For a general local precision shearlet, without specifying the explicit functions used, we see that the condition (4.5) is fulfilled under the following circumstances.

Lemma 4.6. *Let* $\psi = \psi_1 \psi_2$ *be a local precision shearlet, where* ψ_1 *has* q *vanishing moments. Condition (4.5) is fulfilled if and only if* $q = 1$.

Proof. Since ψ_1 is a function point-symmetric to $(0,0)$, the function $\psi_1 x_1$ is axis-symmetric to $x_1 = 0$. Therefore, we have

$$
\int_{-\infty}^{0} \psi_1(x_1)\, x_1 \mathrm{d}x_1 \;=\; \int_{0}^{\infty} \psi_1(x_1)\, x_1 \mathrm{d}x_1
$$
$$
=\; \frac{1}{2} \int_{\mathbb{R}} \psi_1(x_1)\, x_1 \mathrm{d}x_1.
$$

For $q > 1$ we have

$$
\int_{\mathbb{R}} \psi_1(x_1)\, x_1 \mathrm{d}x_1 = 0
$$

and for $q = 1$

$$
\int_{-\infty}^{0} \psi_1(x_1)\, x_1 \mathrm{d}x_1 \neq 0.
$$

This completes the proof.   □

Intuitively, the usage of $p = 1$, i.e. only one vanishing moment, entails that we have only one local extremum in the interval $[-b_1, 0]$. Thus, it is clear that we have $\int_{-b_1}^{0} \psi_1(x_1) x_1 \mathrm{d}x_1 \neq 0$.

The choice of $q = 1$ contradicts the assumptions of Theorem 3.14 for providing a frame for $L^2(\mathbb{R}^2)$. As described earlier, Theorem 3.14 implies that sufficient vanishing moments of ψ are needed. For example, for spline shearlets we need at least $q = 5$, since we have $q \geq \beta'$ with $\beta' \geq \beta + \gamma$, $\beta > 0$ and $\gamma > 2(\beta + 2)$. Therefore, with $q = 1$ we cannot apply Theorem 3.14 to get a frame for $L^2(\mathbb{R}^2)$ with a regular shearlet system $SH(\psi, \Gamma_R)$ or Corollary 3.20 for our shearlet system $\mathcal{LPSH}(\Phi, \Psi, \tilde{\Psi})$. Still, one can choose p and q such that $\mathcal{LPSH}(\Phi, \Psi, \tilde{\Psi})$ forms a frame for $L^2(\mathbb{R}^2)$ according to Corollary 3.20 and such that at least the edge characterization results for a general compactly supported shearlet from 2.21 can be obtained. However, in Section 5.5 we will show that practically the usage of $q = 1$ delivers the best results for pedestrian detection.

Figure 4.4 shows exemplary edge detections of an image of the BSDS500[1] data set provided by the Shearlet Cascade Edge Detection (SCED) algorithm of [34] using different types of shearlets. Algorithm 4.1 presents the SCED algorithm slightly adapted to our shearlet system definition. First, the shearlet energy E is estimated at the finest scale. In the following, all points which are likely to be noise are corrected by taking the energy value at a coarser scale. Finally, a non maximum suppression based on the estimated orientation O and a thresholding are performed. Clearly it can be seen that the local precision shearlet transform provides much more accurate edge estimates than the original FFST and the adaption of Duval-Poo et al. [34]. Correspondingly, we will show in Section 5.5 that the usage of the LPST yields significant improved results for pedestrian detection in comparison to other available shearlet frameworks.

Algorithm 4.1 Shearlet Cascade Edge Detection (SCED) algorithm.

Input: f: input image, $pShear$: shearlet feature parameters containing j_0: number of shearlet scales, ηs: number of shearlets (directions) per scale, B: boundary of mother shearlet support, α: degree of anisotropy, t: parameter for thresholding shearlet coefficients.

Output: E: edge image.

1: **procedure** SCED$(f, pShear)$
2:     $\Psi = $ getShearlets$(B, j_0, \eta s, \alpha)$;
3:     $\mathcal{LPST} = $ locPrcShearletTransform(f, Ψ);
4:     **for all** $m \in f$ **do**
5:         $E(m) = \sqrt{\sum_k (\mathcal{LPST}(j_0 - 1, k, m))^2}$;
6:         $O(m) = \arg\max_k |\mathcal{LPST}(j_0 - 1, k, m)|$;
7:     **end for**
8:     **for** $j = j_0 - 2$ **to** 0 **do**
9:         **for all** $m \in f$ **do**
10:             $e_j(m) = \sqrt{\sum_k (\mathcal{LPST}(j, k, m))^2}$;
11:             $E(m) = \begin{cases} E(m) & \text{if } E(m) \leq e_j(m) \\ e_j(m) & \text{if } E(m) > e_j(m) \end{cases}$
12:         **end for**
13:     **end for**
14:     $E = $ nonMaxSup(E, O);
15:     $E = $ thresholding(E, t);
16:     **return** E;
17: **end procedure**

Furthermore, in Figure 4.4c we illustrate the abovementioned edge artifacts which occur if one uses a shearlet component in x_1 direction with oscillations, for example a B-spline wavelet of order $p \geq 3$. While we can see a clear edge when using the spline shearlet of order $p = 5$

---

[1]Berkeley Segmentation Data Set and Benchmarks 500

with $q = 1$, the replacement of ψ_1 by the B-spline wavelet from (3.8) of order $p = 5$ introduces *shadow edges* nearby the actual edge. These shadow edges are produced by the oscillations of the applied wavelet. Thus, the requirement in Theorems 4.2-4.5 that a local precision shearlet shall have only 1 vanishing moment for a characterization of edge points is in harmony with the observations in Figure 4.4c.

4.3   Conclusion

In this chapter, we studied the capability of local precision shearlets to detect edges in images from a theoretical point of view. We used the setup of Guo and Labate [53] where the image function f is modeled as the characteristic function χ_R of a bounded domain $R \subset \mathbb{R}^2$ with piecewise smooth boundary ∂R. In contrast to the existing literature, we considered a general degree of anisotropy $\alpha \in [1/2, 1)$ during our examination.

Utilizing our shearlets, we showed that edge points $p \in \partial R$ can be characterized by the decay rates of the shearlet transform for decreasing scales. More precisely, we obtained that $\mathcal{SH}_\psi \chi_R (a, s, p)$ decays for $a \to 0$ with $\mathcal{O}(a^{\frac{3-\alpha}{2}})$ if the shear parameter s does not correspond to the normal direction at $p \in \partial R$. If s corresponds to the normal direction $\mathcal{SH}_\psi \chi_R (a, s, p)$ decays with $\mathcal{O}(a^{\frac{1+\alpha}{2}})$. For the special case of $\alpha = 1/2$, we showed that the type of the edge points can be determined by the limit of the shearlet transform.

In order to obtain our theoretical results, we showed that a shearlet mother function with just one vanishing moment is required. Finally, we illustrated that this finding is in harmony with the observed behavior in practice. In the practical application of an edge detection algorithm, the utilization of shearlets with a higher number of vanishing moments resulted in less precise edge detections.

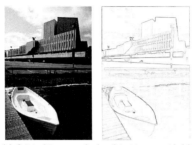

(a) Original image and edge detections provided by
the local precision shearlet transform using the spline
shearlet with $p = 5$ and $q = 1$.

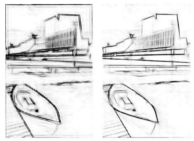

(b) Edge detection using different shearlet frameworks: FFST using Meyer wavelet (left) [60, 61] and
Mallat wavelet (right) [34].

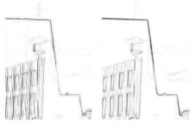

(c) Edge artifacts using the B-spline wavelet from
(3.8) of order $p = 5$ (left) in comparison to edges
provided by the spline shearlet with $p = 5$ and $q = 1$
(right).

Figure 4.4: Edge detection example using different types of shearlets.

"Experience without theory is blind, but theory without experience is mere intellectual play."

Immanuel Kant

5

Pedestrian Detection using Shearlet Features

In this chapter, we investigate if shearlets can provide the framework for generating the best to date hand-crafted features for pedestrian detection. Schwartz et al. [111] defined a simple image feature based on shearlet coefficients and applied it to texture classification and face identification. For pedestrian detection, we define more complex feature types based on the shearlet transform. Given the LPST defined in Section 3.2, we are able to describe the shearlet features considered. Moreover, this chapter will provide the general detection algorithm and the practical realization of it.

Finally, experimental results using shearlet features measured on the Caltech data set [30] are given. Here, we compare the detection quality of gradient features to our approach showing that shearlet features are a promising and outperforming alternative. Especially, we show the improvement over the LDCF++ detector [99] which is currently the best performing detector in the Caltech benchmark using hand-crafted features. Parts of this chapter have been published in [107].

5.1 Pedestrian Detection using Hand-crafted Features

The classical method for the detection of pedestrians in images uses predefined, also called *hand-crafted*, image features. These features are computed for a detection window which is slid over the image such that a *classifier* can decide if the detection window contains a pedestrian or not. Different types of classifiers are used in the current algorithms for pedestrian detection, e.g. Support Vector Machines (SVM) and AdaBoost. However, Benenson et al. [8] find that no classifier type is better suited for pedestrian detection than another. Therefore, a main focus is on the informative content of the image features. The more meaningful the features, the higher is the quality of the detection algorithm.

A breakthrough in the development of hand-crafted image features has been achieved by Dalal and Triggs [22] with their definition of *Histogram of Oriented Gradients (HOG)* features. Here, the image is divided into spatial cells and a histogram of gradient directions is computed over the pixels of the corresponding cell. These local histograms are accumulated and normalized

compute channels   aggregate   vectorize   apply boosted trees

(a) Feature computation and classification for a detection window.

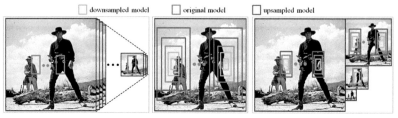

☐ downsampled model   ☐ original model   ☐ upsampled model

(b) Possible pipelines of the sliding window approach for a multiscale pedestrian detection.

Figure 5.1: Illustration of the classical method for pedestrian detection [27, 28].

over a block scheme. Dollár et al. [27, 28] built up on this feature type for their *Aggregated Channel Features (ACF)* detector. This detector also uses gradient histograms but with a different computation method. In addition, it uses gradient magnitude as features. The source code of the ACF detector is available online [26]. According to Benenson et al. [8] in 2014, all best performing pedestrian detection algorithms to that time used hand-crafted features based on HOG or "HOG-like" features, which may encode richer information from the original feature data [7, 96, 99, 105, 128, 130].

The classical method for pedestrian detection is illustrated in Figure 5.1 [27, 28] by the example of the ACF detector. Per detection window, an image transformation is computed. The result of such a transformation is also called *channel*. The image features are computed by aggregation of these channels. Then, the computed features are vectorized and a classifier is applied for the determination to which class the detection window belongs. The ACF detector uses an AdaBoost classifier with boosted decision trees, which is described in [5].

For a detection of pedestrians of different scales, there are different possible pipelines for applying the sliding window approach. In the standard pipeline, a dense image pyramid is created and the detection window is slid over each image of the pyramid with a fixed size. If shift and scale invariant features are used, like done by Viola and Jones [118], a second option is to use a classifier pyramid. This means that a classifier can be placed at any location and any scale such that no image pyramid needs to be created. Finally, the ACF detector [27, 28] uses a hybrid approach of constructing an image pyramid with one octave between two consecutive images. In the following, detector outputs are approximated within half an octave of each pyramid level.

Currently, pedestrian detection algorithms utilizing CNN models [9, 11, 12, 32, 84, 101, 117, 127] outperform detectors using hand-crafted features, see the Caltech Pedestrian Detection Benchmark results online available under http://www.vision.caltech.edu/Image_Datasets/ CaltechPedestrians. We will deal with the capability of shearlets to improve CNN approaches in Chapter 6.

5.2   Shearlet Image Features

We consider digital images in $\mathbb{R}^{M \times N}$ as functions $f \in L^2(\mathbb{R}^2)$ sampled on the grid $\mathcal{G}$ which is given by $\mathcal{G} := \{(cm_1, cm_2) : m_1 = -\lfloor M/2 \rfloor, \ldots, \lceil M/2 \rceil - 1, m_2 = -\lfloor N/2 \rfloor, \ldots, \lceil N/2 \rceil - 1\}$ with $c > 0$. Similarly to the procedure for the computation of HOG [22] or HOG-like [28, 29] features, the image f is partitioned into quadratic *patches* $\mathcal{P}_l$ of predefined size $\zeta \times \zeta \in \mathbb{N}^2$ with localization $l \in \mathcal{L} := \{(l_1, l_2) : l_1 = 1, \ldots, L_1, l_2 = 1, \ldots, L_2\}$, $L_1 = \lfloor M/\zeta \rfloor$ and $L_2 = \lfloor N/\zeta \rfloor$. For each image patch, different types of features are computed based on the shearlet transform. For simplification we set

$$C_{j,k}(m) := \left| \mathcal{LPST}_{\phi, \psi, \tilde{\psi}}(f)(j, k, m) \right| \tag{5.1}$$

for the absolute value of the result of the shearlet transform, namely the *shearlet coefficient*, at pixel $m \in \mathcal{G}$ given scale $j \in \{0, \ldots, j_0 - 1\}$ and shear $k = -\tilde{\eta}_j, \ldots \tilde{\eta}_j$. Since we use vertical and horizontal shearlets, we get two coefficients for each shearing parameter which we denote by $C_{j,k}^1(m)$ and $C_{j,k}^2(m)$. Furthermore, we do not consider the coefficients obtained with the generating function ϕ for our features. Thus, we have $C_{j,k}(m) = (C_{j,k}^1(m), C_{j,k}^2(m))$.

The first feature we introduce, called *shearlet magnitude*, provides orientation independent edge information in a given patch. The provided information is similar to a gradient magnitude, used by the ACF detector [27, 28], but the representation is more sparse due to the sparse image representation provided by the shearlet framework. The feature is defined by taking the normalized sum of shearlet coefficients over all shears in a scale j averaged over all pixels in a patch $\mathcal{P}_l$. This averaging procedure results in a downsampling which is also called *feature pooling*.

We define the shearlet magnitude by

$$M_j(m) := \sum_{i=1}^{2} \sum_{k=-\tilde{\eta}_j}^{\tilde{\eta}_j} C_{j,k}^i(m). \tag{5.2}$$

We perform a local normalization by

$$\widetilde{M}_j(m) := \frac{M_j(m)}{\left(\bar{M}_j(m) + \varepsilon \right)}, \quad \varepsilon > 0, \tag{5.3}$$

with $\varepsilon > 0$ and

$$\bar{M}_j(m) := \sum_{p_1 = -R}^{R} \sum_{p_2 = -R}^{R} w_{p_1, p_2} M_j(m + (p_1, p_2))$$

being the weighted average of the shearlet magnitude in an image area with center m and radius R, i.e. size $(2R+1) \times (2R+1)$. The weights w_{p_1, p_2} are set to fulfill the conditions $0 \le w_{p_1, p_2} \le 1$ for all $p_1, p_2 \in \{-R, \ldots R\}$ and $\sum_{p_1 = -R}^{R} \sum_{p_2 = -R}^{R} w_{p_1, p_2} = 1$. This normalization scheme is adopted from the ACF detector [27, 28]. Using this notation, the shearlet magnitude feature for patch $\mathcal{P}_l$ is computed by

$$\mathcal{SM}_j(l) := \frac{1}{\zeta^2} \sum_{m \in \mathcal{P}_l} \widetilde{M}_j(m). \tag{5.4}$$

The procedure for calculating the shearlet magnitude feature is illustrated in Figure 5.2 for a sample image of the Caltech test data set with a fixed parameter j. A colormap has been applied to visualize data values from high to low by blue through white to red.

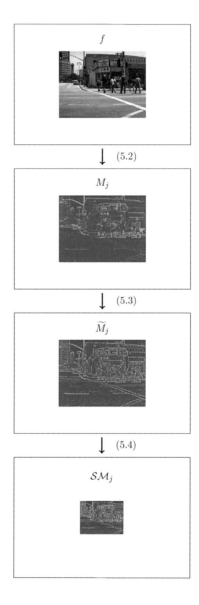

Figure 5.2: Procedure for calculating the shearlet magnitude feature for a fixed parameter j with an exemplary patch size of 2×2.

To also incorporate the information about dominant orientations in a patch, a *shearlet histogram* is used as second type of feature. Given a fixed scale j, each pixel m of a patch P_l is assigned to the direction, i.e. its shear index, with the largest absolute value of the corresponding shearlet coefficients. This operation can be expressed by

$$O_j\left(m\right) := \underset{(i,k)\in\{1,2\}\times\{-\bar{\eta}_j,\ldots\bar{\eta}_j\}}{\arg\max}\ C_{j,k}^i\left(m\right).$$

Then, the shearlet histogram for patch P_l is set up by $\mathcal{SH}_{j,k}(l) = (\mathcal{SH}_{j,k}^1(l), \mathcal{SH}_{j,k}^2(l))$ and

$$\mathcal{SH}_{j,k}^i\left(l\right) := \frac{1}{\zeta^2}\sum_{m\in P_l}\widetilde{M}_j\left(m\right)\chi_{\{(i,k)\}}\left(O_j\left(m\right)\right), \tag{5.5}$$

where χ is the characteristic function. The informational content of it is similar to a gradient histogram used in [27, 28]. Note that we get two features for every k. An exception is the diagonal case, i.e. η_j divisible by 4, where we get one value for $|k| = 1$ since only one of two shearlets is kept.

These shearlet features share the property of directly providing multi-scale information empowered by the shearlet framework. In that way, it is possible to detect structures from fine to coarse scales at the same time and to locate which type of structure is contained in which image area via the patch partitioning. Figure 5.3 shows a visualization of the shearlet features averaged over a set of pedestrian images for a fine scale j, 6 directions ($\eta_j = 6$), and a trivial patch size of 1×1 ($\zeta = 1$). The shearlet magnitude feature in the top row gives a good characterization of the pedestrian's silhouette in total. The shearlet histogram feature in the bottom row of Figure 5.3 shows a strong response to certain body parts depending on the analyzed direction. For example the first feature component, using vertical directed shearlets, responds heavily to the side of the pedestrian's silhouette. Feature components using sloped directions show a good response to shoulder parts.

In comparison to the feature descriptor described by Schwartz et al. [111], these two kinds of features provide much richer and more stable information about an object's structure. In [111], shearlet features are defined simply by the sum of absolute values of the shearlet transform over all image pixels m for each scale j and shear k. The corresponding feature $H_j\left(k\right)$ is computed by

$$H_j\left(k\right) := \sum_{m_1=-\lfloor M/2\rfloor}^{\lceil M/2\rceil-1}\sum_{m_2=-\lfloor N/2\rfloor}^{\lceil N/2\rceil-1}\left|\mathcal{SH}_\psi\left(f\right)\left(j,k,m\right)\right|. \tag{5.6}$$

This corresponds to a shearlet coefficient feature without any normalization or patch localization. As we will show in Section 5.5.2 using this feature even with normalization and patch localization yields inferior detection results.

5.3   Shearlet Filterbank

In addition to shearlet features computed on input images, we consider an intermediate filtering layer between original image feature computation and classification as described in [130]. After computation of the original feature maps, the feature vector used for classification is built by a sum pooling over rectangular regions. According to [130], this sum pooling can be rewritten as a convolution with a filterbank, where one filter is used per rectangular shape. Following, a single value of the convolution's response map is used as feature. This procedure is illustrated in Figure 5.4.

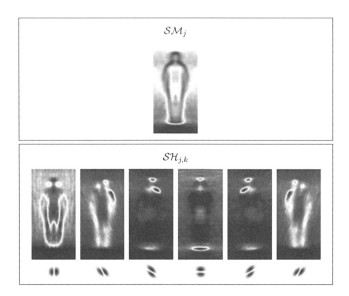

Figure 5.3: Visualization of shearlet features for a fixed scale and 6 orientations averaged over a set of pedestrian sample images. Top: shearlet magnitude, bottom: shearlet histogram.

In our case, the input features for the intermediate layer can be either color, gradient or shearlet features. As filters, we use shearlets from our local precision shearlet system. Since this intermediate layer is executed as a convolution with these shearlet filters, one can regard the result as shearlet coefficients of image features. Given a *feature map* $\mathcal{F} \in \mathbb{R}^{\widetilde{M} \times \widetilde{N} \times D}$ of size $\widetilde{M} \times \widetilde{N}$ and depth $D \in \mathbb{N}$, we have the *shearlet filterbank feature*

$$\mathcal{SF}_{j,k}(l) := \frac{1}{\tilde{\zeta}^2} \sum_{m \in \widetilde{\mathcal{P}}_l} \mathcal{LPST}_{\phi,\psi,\tilde{\psi}}(\mathcal{F})(j,k,m) \tag{5.7}$$

for a scale j, a shear k and a feature map patch $\widetilde{\mathcal{P}}_l$ of size $\tilde{\zeta} \times \tilde{\zeta}$. In addition to the resulting filtered features, we preserve the original low level image features in the final feature space.

The setup of the shearlet filterbank is inspired by the filterbank of *RotatedFilters* [129]. For color and magnitude features, we use only the horizontal and vertical shearlets. For each directional feature, such as a gradient or a shearlet histogram, we use the shearlet corresponding to the feature direction and its counterpart from the other spatial cone. It is worth mentioning that such a shearlet filtering of gradient histograms is only possible with shear parameters set like in the local precision shearlet system. Using other shearlet systems, the most shearlet directions are not in correspondence to the directions of the gradient histogram features. Only the directional selection as applied in a local precision shearlet system, guarantees this correspondence. Summarizing, for each image feature, we have a pair of shearlet filters per scale. Furthermore, we add a simple 1×1 square to the filterbank in order to preserve the original image features in the final feature map. That means, we have a filtered feature space size of $\sigma_f = \sigma_o \cdot (2j_0 + 1)$, where σ_o is the size of the original feature space. Figure 5.5 illustrates its filters for one scale

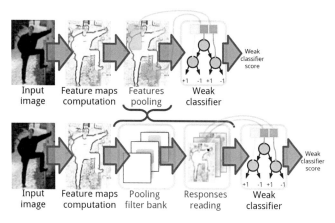

Figure 5.4: Intermediate layer filtering image features [130].

Figure 5.5: Illustration of the shearlet filterbank. The top row contains the low level image feature maps. L, U and V correspond to the color channels of the image and $|\cdot|$ denotes a magnitude feature. The rotated bars represent directional histogram features. The middle and the bottom row show the pair of shearlet filters for each of these feature maps.

and the connection to the original image feature maps. Here, LUV features are used for color information. The LUV color space consists of three channels: L represents the luminance, U and V the color type.

5.4　Implementation Details

The detection algorithm is based on a sliding window approach, where a detection window of fixed size is slid over the image followed by a binary classification if a pedestrian is contained in the window or not. This procedure is applied on different scales of the input image to allow for different pedestrian sizes due to physical differences or differing distances to the camera. This is a typical and widely used approach for pedestrian detection, see [31]. For binary classification, a decision function is needed, which is typically carried out by a machine learning algorithm. Based on feature vectors from a training data set containing positive and negative sample images of the size of the detection window, the machine learning algorithm learns which feature values belong to which class: "pedestrian" or "no pedestrian". During the application of pedestrian detection, the classifier then compares the measured features from the detection window to the learned values. Besides the binary classification if the detection window contains a pedestrian or not, the classifier also returns a classification score. To avoid nearby multiple detections of the same pedestrian only the detection with the maximum score is preserved if some detection windows are overlapping by a predefined threshold. This technique is known as non maximum suppression (NMS). As a general detection framework, we use an adaption of the ACF detector by Dollár et al. [27, 28]. Especially, we use the same AdaBoost classifier which is described in [5]. In this way, we are able to directly compare the feature performance avoiding potential side effects because of a different classifier. For a d-dimensional feature vector $x \in \mathbb{R}^d$, the output of classifier we use is expressed by

$$H(x) = \sum_{t=1}^{T} \lambda_t h_t(x),$$

with $\lambda_t \in \mathbb{R}$ and weak learners h_t, which are optimized during the training process. The weak learners are shallow decision trees with

$$h_t(x) = p_t \mathrm{sgn}(x_{d_t} - \tau_t),$$

with a polarity $p_t \in \{-1, 1\}$, a feature index $d_t \in \{1, \dots, d\}$ and a threshold τ_t. We will describe the parametrization of the classifier used in our experiments in Section 5.5.1.

The first variant of our detector, called *shearFtrs-v1* detector, is illustrated in Algorithm 5.1. In this variant, we only compute low level image features based on the shearlet transform. The application of a shearlet filterbank is described separately later on.

At first, the scales for scaling the input image are calculated such that the detection window still fits into the smallest downscaled image. In addition, the local precision shearlet filters are set up according to (3.1)-(3.11). Since the shearlet filters are not depending on the input image or its size, the computation of them does not have to be done per image and can be outsourced in case of multiple input images. Then, the LPST is calculated for each scaled image by (3.14). Concerning input image type, we use images converted to LUV color space. To obtain the shearlet coefficients of a LUV color image, we compute the shearlet coefficients for each color channel separately and take for each pixel the shearlet coefficient with the maximum absolute value.

The input image is not smoothed in order to denoise it before the shearlet transform is computed. In our algorithm, we reduce the influence of noise from the input image by a shearlet thresholding approach as described by Easley and Labate [36]. In this procedure, shearlet coefficients whose absolute values fall below the predefined threshold *tShear* are set to zero. Given the result, one can use the inversion formula of the shearlet transform in order to restore the image. However, we are not interested in the output of a denoised image. Therefore, we use the modified shearlet coefficients to calculate the shearlet features according to (5.1)-(5.4). Corresponding to the normalization scheme of the ACF detector [27], the weights for the normalization of the features are set as constants by the values of a triangle filter.

The parameters for setting up the shearlet filters as well as for computing the LPST and shearlet features include the number of shearlet scales j_0, the number of shears η_j per scale j, the size of the mother shearlet support specified by its boundary B, the degree of anisotropy α, the normalization radius R and the patch size $\zeta \times \zeta$. The features are formatted such that the patches are contained in the first two dimensions and the scales and shears in the third dimension. In other words, we have $\mathcal{SM} \in \mathbb{R}^{L_1 \times L_2 \times j_0}$ and $\mathcal{SH} \in \mathbb{R}^{L_1 \times L_2 \times j_0 \eta}$ with $\eta = \sum_{j=0}^{j_0-1} \eta_j$.

Subsequently, the features are concatenated along the fourth dimension, smoothed and given to the sliding window approach as as feature map input together with a trained classifier and a predefined stride. The AdaBoost classifier requires all features of a detection window being in a single feature vector. Therefore, the computed features are vectorized before applying the

---

Algorithm 5.1 Procedure of the shearFtrs-v1 detector.

Input: f: input image, C: classifier, Ds: dimensions of detection window, s: stride for sliding window approach, $tNms$: threshold for non maximum suppression, $pShear$: shearlet feature parameters containing j_0: number of shearlet scales, ηs: number of shearlets (directions) per scale, B: boundary of mother shearlet support, α: degree of anisotropy, $tShear$: parameter for thresholding shearlet coefficients, R: normalization radius, ζ: patch size.

Output: bbs: bounding boxes of detections.

1: **procedure** SF-V1_DETECT($f, C, Ds, s, tNms, pShear$)
2:     $scls = \texttt{getScales}(f, Ds)$;
3:     $\Psi = \texttt{getShearlets}(B, j_0, \eta s, \alpha)$;
4:     **for** $i = 1$ **to** $\texttt{length}(scls)$ **do**
5:         $g = \texttt{scaleImage}(f, scls_i)$;
6:         $\mathcal{LPST} = \texttt{locPrcShearletTransform}(g, \Psi)$;
7:         $\mathcal{LPST} = \texttt{shearletThrsh}(\mathcal{LPST}, tShear)$;
8:         $C = |\mathcal{LPST}|$;
9:         **for** $j = 0$ **to** $j_0 - 1$ **do**
10:             $M_j = \sum_{k \in \Lambda} C_{j,k}$;
11:             $\widetilde{M}_j = \texttt{normalize}(M_j, R)$;
12:         **end for**
13:         **for all** $l \in \mathcal{L}$ **do**
14:             **for** $j = 0$ **to** $j_0 - 1$ **do**
15:                 **for all** $k \in \{-\tilde{\eta}_j, \ldots, \tilde{\eta}_j\}$ **do**
16:                     $\mathcal{SH}^1_{j,k}(l) = {}^1/\varsigma^2 \sum_{m \in \mathcal{P}_l} \widetilde{M}_j(m) \chi_{(1,k)}(O_j(m))$;
17:                     $\mathcal{SH}^2_{j,k}(l) = {}^1/\varsigma^2 \sum_{m \in \mathcal{P}_l} \widetilde{M}_j(m) \chi_{(2,k)}(O_j(m))$;
18:                 **end for**
19:                 $\mathcal{SM}_j(l) = {}^1/\varsigma^2 \sum_{m \in \mathcal{P}_l} \widetilde{M}_j$;
20:             **end for**
21:         **end for**
22:         $\mathcal{F} = \texttt{concatenate}((\mathcal{SM}, \mathcal{SH}), 4)$;
23:         $\mathcal{F} = \texttt{smooth}(F)$;
24:         $bbs_i = \texttt{applySlidingWindow}(\mathcal{F}, C, Ds, s)$;
25:     **end for**
26:     $bbs = \texttt{transformCoords}(bbs_1, \ldots, bbs_n, scls)$;
27:     $bbs = \texttt{applyNms}(bbs, tNms)$;
28:     **return** bbs;
29: **end procedure**

---

classifier. After classification the coordinates of detected pedestrians, also called *bounding boxes*, obtained from scaled images have to be transformed to coordinates of the original image. Finally, the already mentioned non maximum suppression is applied to provide the final bounding boxes of detected pedestrians.

Our second variant named *shearFtrs-v2* detector, i.e. the application of a shearlet filterbank, is illustrated in Algorithm 5.2. After computation of image scales and shearlets as in Algorithm 5.1, the shearlet filterbank is set up corresponding to the basic image feature types specified in the parameter $pFtr$. These basic image features can contain color, gradient or shearlet features. Next, the corresponding feature map is computed for each image scale. For each element of the shearlet filterbank, a convolution of the feature map with the corresponding filter element is performed. This is equivalent to the computation in (5.7). As in Algorithm 5.1, the feature map is given the sliding window approach as input to compute the bounding boxes of detected pedestrians. Figure 5.6 shows exemplary detection results of a test image of the Caltech data set [30] using the shearFtrs-v2 detector. Corresponding bounding boxes of detected pedestrians are colored in green.

Algorithm 5.2 Procedure of the shearFtrs-v2 detector, i.e. the filterbank variant.

Input: f: input image, $pFtr$: feature map parameters, C: classifier, Ds: dimensions of detection window, s: stride for sliding window approach, $tNms$: threshold for non-maximum suppression, $pShear$: shearlet feature parameters containing j_0: number of shearlet scales, ηs: number of shearlets (directions) per scale, B: boundary of mother shearlet support, α: degree of anisotropy, $\tilde{\zeta}$: patch size.

Output: bbs: bounding boxes of detections.

 1: **procedure** SF-V2_DETECT($f, C, Ds, s, tNms, pFtr, pShear$)
 2:     $scls = \texttt{getScales}(f, Ds)$;
 3:     $\Psi = \texttt{getShearlets}(B, j_0, \eta s, \alpha)$;
 4:     $fs = \texttt{shearFilters}(\Psi)$;
 5:     **for** $i = 1$ **to** $\texttt{length}(scls)$ **do**
 6:         $g = \texttt{scaleImage}(f, scls_i)$;
 7:         $\mathcal{F} = \texttt{computeFtrMap}(g, pFtr)$;
 8:         **for** $j = 1$ **to** $\texttt{numOfElements}(fs)$ **do**
 9:             $\tilde{\mathcal{F}} = \texttt{conv2}(F, fs_j)$;
10:             $\mathcal{SF}_j(l) = 1/\tilde{\zeta}^2 \sum_{m \in \tilde{\mathcal{P}}_l} \tilde{F}(m)$;
11:         **end for**
12:         $bbs_i = \texttt{applySlidingWindow}(\mathcal{SF}, C, Ds, s)$;
13:     **end for**
14:     $bbs = \texttt{transformCoords}(bbs_1, \ldots, bbs_n, scls)$;
15:     $bbs = \texttt{applyNms}(bbs, tNms)$;
16:     **return** bbs;
17: **end procedure**

5.5   Experiments

In this section, we show experimental results for pedestrian detection using shearlet features and demonstrate their ability to provide better detection rates than gradient features. We evaluate the detection quality by plotting *Receiver Operating Characteristics (ROC)* curves, which show a tradeoff of the detection or miss rate against the false positives per image (fppi). All evaluations

Figure 5.6: Example output of the shearFtrs-v2 detector. Detected pedestrians are represented by green colored bounding boxes.

have been done on *reasonable experiment* of the Caltech benchmark [30, 31], which is widely used for pedestrian detection benchmarks [8, 31, 129]. In this experiment, all pedestrians of height 50 pixels and taller are evaluated, who are only slightly occluded, i.e. visible for at least 65%.

As indicated before, we are using the same detection framework and AdaBoost classifier as the ACF detector [27, 28]. All currently top performing hand-crafted feature detectors are using this framework [99, 129, 130]. This way we are able to directly compare the feature performance avoiding potential side effects because of a different classifier. Besides the comparison to other hand-crafted features, we also show the impact of parametrization of the feature computation. Furthermore, we show detection results using different shearlet implementations such as FFST [61] and ShearLab 3D [77].

Figure 5.7 shows ROC plots for the best achieved result using shearlet features and all other top performing algorithms on the Caltech data set. The first algorithm not using a CNN framework is the LDCF++ detector [99]. This detector is a modification of the LDCF detector [96], which uses ACF features and an intermediate filtering layer to remove correlations in local neighborhoods. In the term of LDCF++ detector, the first '+' stands for adapted boosting and augmentation parameters and the second '+' for contextual analysis. A location prior model capturing spatial context is used to recalculate the object score of the original hand-crafted feature detector. With 14.75%, the miss rate using shearlet features is below the one of the LDCF++ detector having a miss rate of 14.98%. Comparing the pure feature detector results by excluding the contextual analysis, the LDCF+ detector achieves a miss rate of 15.40%. Thus, with a miss rate of 14.75%, the shearFtrs-v2 detector significantly outperforms LDCF+. Although the usage of shearlets for computing hand-crafted features shows an improvement, still the detection results of CNN approaches can not be reached. This is another indication that hand-crafted feature detectors will not be able to compete with CNNs, nor by optimization of classifier parameters as in [99] nor by definition of better features as in our work.

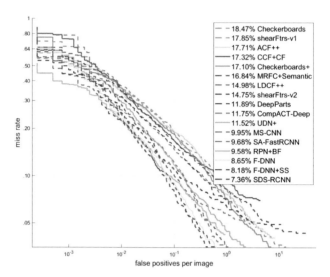

Figure 5.7: ROC plots for currently top performing pedestrian detection algorithms on the Caltech data set.

5.5.1 Detection Framework Settings

Like the ACF detector [27, 28], we use detection windows of size 128×64 and a stride of 4 for the sliding window procedure. We also use the same values for the number of image scales per octave by 8 and the threshold for non maximum suppression of 65%.

We use the same classifier set up as in [99]. In this way, we have a fair comparison to the gradient features of the ACF+ detector and the currently best performing hand-crafted features of the LDCF+ detector. We use 4 stages of bootstrapping for training the AdaBoost classifier on the Caltech data set with an increasing number of weak learners and a maximum of $4,096$ at the final stage. Furthermore, we use decision trees with a maximum depth of 5 as well as $48,996$ of positive and $200,000$ of negative sample images. The total number of positive training samples is achieved by a data augmentation method as described in [99]. We sample every 4-th frame of the training data set video sequences and scale the original samples by a factor of 1.1 in horizontal, vertical and both directions.

5.5.2 Image Feature Experiments

First, we evaluate the performance of the shearFtrs-v1 detector using shearlet features computed directly from the original image. Before stating the comparison results, we give an overview of the default parametrization which has been used to obtain the best results. We use 3 scales with 12 shears per scale. The mother shearlet is chosen to be a spline shearlet of order $m = 3$ and 1-st derivative. We set the patch size parameter to $\zeta = 4$, leading to a total number of $32 \cdot 16 \cdot 3 \cdot (12 + 1) = 19,968$ shearlet features. The radius for normalizing the shearlet magnitude during feature computation is set to $R = 11$. In addition, we use LUV color information as feature like the ACF detector [27, 28].

On the Caltech data set, we achieve a log-average miss rate of 17.85%. The corresponding ROC plot is shown in Figure 5.7 with denotation shearFtrs-v1. In comparison to gradient features, Ohn-Bar and Trivedi [99] reported a log-average miss rate of 20.69% with their ACF+ detector as best known ACF result. Therefore, the usage of shearlet image features yields a significant improvement in detection rates. The improvement amounts to such an extent that the detector can compete with more enhanced hand-crafted feature approaches such as a filterbank application, e.g. the Checkerboard detectors [130].

We now investigate the effect of key parameters on the detection quality. Especially of interest are parameters used for computation of the shearlet features defined in Section 5.2. We concentrate on the results when changing values of one parameter and letting the other parameters fixed to the best performing value. In general, we found consistent results if we change more than one parameter at a time. Consider that we change one parameter to a value showing decreased detection performance in a single parameter test. If we change another parameter also to a worsening value of the corresponding single parameter test, the detection result further decreases.

Shearlet Scales

One major parameter is the shearlet scales used for feature computation. First, we evaluate different numbers of shearlet scales, i.e. $j_0 = 1, 2, 3$. Further scales reduce the support of the corresponding shearlets to such an extent that they are useless for practical application. A larger support of the mother shearlet is enabling larger values of j_0. Unfortunately, this leads to inadequate edge detections at the coarsest scale. Furthermore, we also evaluate the performance of single scales and different combinations of them.

Figure 5.8 shows the ROC plots for the different setups of scales used. While considering only the coarsest scale, the detection quality drops to a log-average miss rate of 29.12%. Including the second scale improves to 22.29% and to 17.85% with inclusion of the third scale. Concerning the evaluation of single scales, we find that the finer the scale, the better is the detection performance. While the first two scales achieve log-average miss rates of 29.12% and 21.24%, considering only the third scale scores 20.59%, which is slightly better than the ACF+ result. Although, the coarsest scale performs significantly worse than the others, it still provides useful information. Sparing the incorporation of it reduces the detection quality from 17.85% to 18.80%.

Shearlets Per Scale

Besides the number of scales, the number of shearlets per scale η_j is a main parameter for the shearlet filter setup. We mainly use a constant number of shearlets across scales to have the same directional analysis on each scale level ranging from 6 to 14 shears per scale. Furthermore, we test the detection performance with the number of shearlets of a regular cone-adapted discrete shearlet system, i.e. 4, 8 and 16 shearlets per scale, and other schemes of increasing numbers of shearlets per scale.

The evaluation of influence of η_j is shown in Figure 5.9. The best performance is achieved with 12 shearlets resulting in a log-average miss rate of 17.85%. A reduction to less shearlets per scale as well as an increase to $\eta_j = 14$ for $j = 0, 1, 2$ results in a significant decrease of detection quality. A consideration of more shears per scale increases the feature space. Thus, it is not

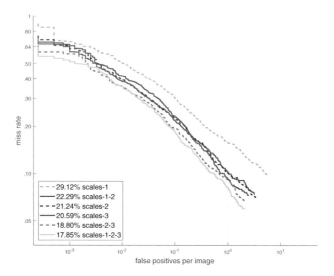

Figure 5.8: Evaluation of number of shearlet scales.

surprising that there is a number of shearlets after which the feature space becomes too large leading to a decreasing classifier performance. On the other hand, using less than 12 shearlets per scale reduces the detection quality significantly since insufficient information is available. Also a utilization of the number of shearlets of a regular cone-adapted discrete shearlet system, with corresponding degree of anisotropy $\alpha = 1/2$ yields drops in detection rates. Finally, variants with other schemes of increasing numbers of shearlets per scale also show minor results.

Degree Of Anisotropy

Next, we analyze the impact of the degree of anisotropy $\alpha \in [1/2, 1)$ on the detection results. More precisely, we use values of $\alpha = 1/2$, $\alpha = 3/4$ and $\alpha = 9/10$.

In Figure 5.10, we show the ROC plots for the tested values. The parameter value of $\alpha = 3/4$ performs best, whereas the value of $\alpha = 1/2$, used in the regular shearlet systems, shows significantly inferior performance with a log-average miss rate of 19.28%. This justifies the use of the general scaling matrix $A_{a,\alpha}$ instead of the standard matrix A_a. Also values close to 1, such as $\alpha = 9/10$, show an inferior performance, e.g. with log-average miss rate of 19.78%.

Different Mother Shearlets and Shearlet Systems

Now, we show the impact of using different kinds of shearlets. First, we analyze the impact of increasing the spline order p and the number of derivatives q of the mother shearlet function. Moreover, we compare the results obtained by using the LPST against the ones achieved with

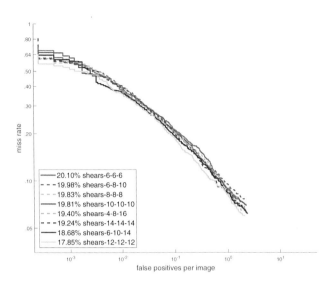

Figure 5.9: Impact of shears per scale.

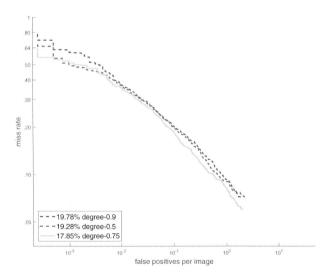

Figure 5.10: Comparison of values for degree of anisotropy.

the FFST [61] and ShearLab 3D [77]. For the application of these toolboxes, we mainly use their standard parameters.

As can be seen in Figure 5.11a, detection rates significantly decrease as we increase either the spline order or the number of derivatives of the underlying B-spline. Increasing the spline order from $p = 3$ to $p = 5$, lowers the log-average miss rate from 17.85% to 21.77%. If we further set the number of derivatives to $q = 3$, detection rates further decrease to a log-average miss rate of 24.79%. This trend proves to be true for even higher values. The usage of $p = 7$ and $q = 5$ performs significantly worse with a log-average miss rate of 27.88%. As explanation for the observed behavior, an increase in the spline order implicates an increase in the support of the shearlet. Furthermore, higher numbers of derivatives lead to more oscillations in the mother function. Both facts result in a less precise localization of edges in images.

Figure 5.11b presents the ROC plots for the shearlet implementations considered. One can see that the LPST outperforms ShearLab 3D and the FFST significantly. ShearLab 3D performs second best with a log-average miss rate of 24.33%. A utilization of the FFST results in a log-average miss rate of 34.82%. As described by Kutyniok et al. [77], the ShearLab 3D implementation also provides functions with compact support in time domain. This can explain why it delivers better results than the FFST. The LPST is more flexible in regards to the number of shearlets per scale since only $\eta_j \in 2\mathbb{N}$ is required. The FFST [61] is bound to $\eta_j = 2^{j+2}$ and ShearLab 3D to $\eta_j = 2^{d_j+2}$ with $d_j \in \mathbb{N}$. Consequently, the best performing parametrization of η_j presented above is not possible for both toolboxes. For ShearLab 3D, we have chosen $d_j = 1$ for $j = 0, 1, 2$, i.e. $\eta_j = 8$, since it corresponds to the next best parametrization from our experiments concerning shearlets per scale, see Figure 5.9. Comparing the results for $\eta_j = 8$, still the LPST performs significantly better than ShearLab 3D. We conclude, that our local precision shearlets can generate more qualitative image features.

It is worth mentioning that gradient features of the ACF+ detector perform significantly better than ShearLab 3D. As stated before, ACF+ achieves a log-average miss rate of 20.69%. That means, we are only able to outperform gradient features with an application of the LPST. However, it is not clear what results can be achieved with more extensive parameter tests when using other shearlet toolboxes.

Feature Type

Finally, we analyze the informative content of each feature type. In other words, we present the detection performance of the shearlet magnitude and the shearlet histogram when they are used as single features. Moreover, we show the result of a feature type according to the feature $H_j(k)$ in (5.6) described by Schwartz et al. [111] as a single feature and the result if we add it to our features.

Figure 5.12 shows the corresponding ROC plots. The best result as a single feature is achieved by the shearlet histogram with a log-average miss rate of 19.48%. The shearlet magnitude feature only achieves a log-average miss rate of 51.92%. A shearlet coefficient feature corresponding to $H_j(k)$ [111], denoted by *shearCoeffs*, results in a detection with log-average miss rate of 35.76%. Adding this feature type to the shearlet magnitude and histogram does not yield an improvement. Detection rates drop to a log-average miss rate of 21.63%.

5.5.3 Filterbank Experiments

Now, we evaluate the performance of our second variant shearFtrs-v2. As described in Section 5.3, we use LUV features with either gradient or shearlet features as input for the inter-

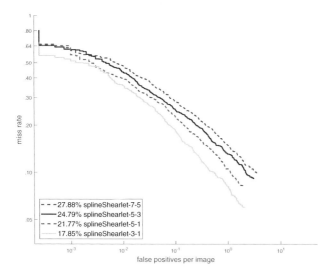

(a) Evaluation of orders and derivatives of spline shearlets.

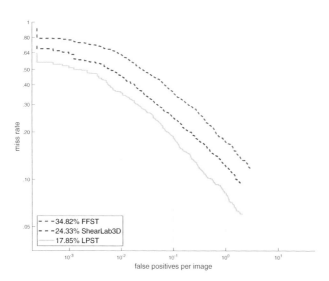

(b) Comparison of LPST to the shearlet frameworks FFST and ShearLab 3D.

Figure 5.11: ROC plots of the shearFtrs-v1 detector evaluating different types of shearlets.

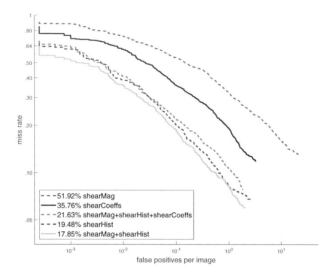

Figure 5.12: Impact of image feature types.

mediate filtering layer. Although shearlet features perform better than basic image features as shown in Section 5.5.2, the filterbank approach shows better results with gradient features as input. To set up the filterbank, we use the standard options of the ACF and ACF+ detectors with 6 orientations for the gradient histogram. Thus, we have 6 shearlets per scale in our filterbank. Furthermore, we use the same characteristics for the generation of the shearlets as in Section 5.5.2. During our experiments, we found equivalent results to our image feature experiments concerning shearlet scales, degree of anisotropy and shearlet mother function parameters. Overall, this setup achieves the best result using shearlet features on the Caltech data set. As stated before, it is also the best known result on the Caltech data set using hand-crafted features, see Figure 5.7.

An explanation for the better performance when using gradient input features is that the utilization of shearlet features significantly increases the feature space. Using gradients with 6 orientations for the gradient histogram entails $35,840$ filtered features. If we use shearlet features with 6 shearlets per scale, the final feature space size enlarges to $86,016$ features and to $107,520$ in case of 8 shearlets per scale. This can lead to an overfitting effect, i.e. the detector can not generalize well from the training data set to test images it has not seen before. Consequently, the detection rates decrease if the feature space size increases. For shearlet input features, we measure a log-average miss rate of 16.35% with 6 shearlets per scale and 17.05% with 8 shearlets per scale.

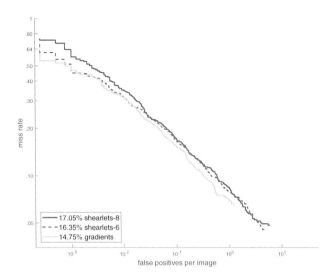

Figure 5.13: Comparison of filterbank results with gradient and shearlet features as input.

5.6 Conclusion

In this chapter, we were investigating the applicability of the shearlet transform to provide meaningful hand-crafted features for pedestrian detection. The idea behind it was that shearlets are able to extract directional information from images, which is widely used in state-of-the-art algorithms. Shearlets have multiple benefits compared to gradients concerning image features or other filters in an intermediate filtering layer. Main advantages are the sparse image representation and the multi-scale framework.

Using these properties, we defined two types of image features based on the shearlet transform, namely the shearlet magnitude and shearlet histogram features. Based on our experimental results on the Caltech data set [30], these features turned out to be very informative. We showed that these features are able to outperform the gradient features of the ACF detector [27], which all best performing hand-crafted features are using as a base. Furthermore, we introduced an application of a shearlet filterbank for an intermediate filtering layer as described in [130]. Using the abovementioned data set, we showed that this filterbank provides performance improvements compared to other known filterbanks leading to the best known results using a hand-crafted feature detector.

A crucial issue for computing high quality shearlet features is the design of the underlying shearlets since edges have to be located very precisely. For that purpose, we introduced local precision shearlets based on compactly supported shearlets in Chapter 3. The corresponding shearlet transform delivers precisely located edge estimates which results in a higher quality of pedestrian detection compared to other shearlet implementations. Furthermore, our flexible shearlet system in regards to number of shearlets per scale provides more control over the feature space size and enables the usage of a shearlet filterbank similar to the RotatedFilters filterbank [129].

With the result that shearlet features can outperform all other current hand-crafted features, it has to be investigated if CNN approaches are subject to performance improvement when integrating shearlets. For example, the subsequent application of a CNN as in [84], which uses pedestrian proposals generated by gradient features, promises an improved detection if one uses shearlet features instead. Furthermore, since the first layer of a CNN can be regarded as to perform extraction of basic features like edges, one can imagine detection performance improvements by an integration of shearlet filters in this layer. Due to the current dominance of CNN approaches in the Caltech benchmark, we investigate this topic in the next chapter.

"Learning never exhausts the mind."
Leonardo da Vinci

6

Deep Learning with Shearlets

Deep learning methods such as *Convolutional Neural Networks (CNNs)* are the base of current best performing algorithms for pedestrian detection. In Chapter 5, we show that we are able to improve hand-crafted feature detectors with shearlet features but that detection algorithms utilizing CNNs still show better detection rates. However, the question arises, if a concentration on pure neural networks is the optimal approach. One may achieve even better results than currently measured if we include insights from hand-crafted feature based pedestrian detection approaches and theoretical frameworks. To this end, we exploit the possibilities to integrate shearlets in CNNs. We aim to use shearlet filters at early convolution layers of a CNN instead of learned ones in order to improve its classification and detection results. The underlying idea is that early CNN layers intuitively perform an edge detection, whereas shearlets theoretically provide optimal filters for this task. Moreover, they can provide a good base for learning filters of deeper layers.

First, we give an introduction about Artificial and Convolutional Neural Networks and the application of the latter for pedestrian detection. Next, we state results of related works concerning the integration of theoretical frameworks in CNNs and describe our own concept and its implementation. Finally, we illustrate the capability of shearlets to improve CNNs in experiments on pedestrian classification and detection.

6.1   Introduction on Neural Networks

First, we will introduce the concept of Artificial and Convolutional Neural Networks and its basic terminology. This introduction is based on the description of Aghdam and Heravi [2].

Artificial Neural Networks (ANNs) consist of connected components called neurons. The basic idea of ANNs is an imitation of biological neurons. A major feature of these networks is the ability of learning to solve specific problems. Furthermore, we will describe the adaptation to CNNs, which has been developed for image processing tasks.

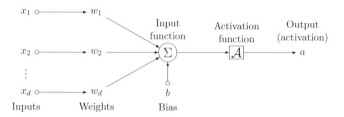

Figure 6.1: Illustration of an artificial neuron

6.1.1 Artificial Neural Networks

An Artificial Neural Network is composed of connected (artificial) neurons. An illustration of such a neuron is given in Figure 6.1. The input of a neuron is calculated by the vector $x \in \mathbb{R}^d$, $d \in \mathbb{N}$, which is multiplied component-wise with weights $w_1, \ldots, w_d$. The weighted input values are then summed up. Optionally, a bias $b \in \mathbb{R}$ is also included during summation. The summed value is then fed into an activation function $\mathcal{A}$ which finally yields the output value $a \in \mathbb{R}$, also called *activation*. There are different types of activation functions, which can be used. They can be chosen for a specific problem to be solved.

According to [2], the special case of a *feedforward neural network* is commonly for computer vision tasks. Especially, convolutional neural networks are a specific type of feedforward networks. Therefore, we will refrain from describing different types of Artificial Neural Networks. A feedforward network is made up of a number of layers containing various numbers of neurons. Neurons of a layer are connected with all neurons of the subsequent layer. For this reason, feedforward neural networks are called to be fully-connected.

The first layer is called *input layer* denoted by I, while the last one is called *output layer* denoted by Z. All layers in between are called *hidden layers* denoted by H_l with $l = 1, 2, \ldots, L \in \mathbb{N}$. The number of neurons in a hidden layer H_l is denoted by d_l. Formally, we set $H_0 := I$ and $H_L := Z$. Figure 6.2 illustrates a simple example of a feedforward neural network consisting of 3 layers in total, thus 1 hidden layer. The dimension of the output vector z is determined by the number of neurons in the output layer.

Now, we define the notation of the variables involved in a neural network which will be used for describing calculation procedures. The weight $w_{i,j}^l$ denotes the connection of the neuron i in layer $l-1$ to the neuron j in layer l. Exemplary, the weight $w_{2,3}^3$ of the third neuron in the second layer to the second neuron in the third layer. A similar notation is used for bias and neuron outputs, i.e. activations. Bias b_j^l is located at neuron j in layer l. Analogously, the activation of neuron j in layer l is denoted by a_j^l. Usually, all neurons in the same layer share the same activation function to compute their activation values. We will mainly use the variables w^l, b^l and a^l, where $w^l \in \mathbb{R}^{d_{l-1} \times d_l}$ contains the weights connecting layers H_{l-1} and H_l as well as $b^l, a^l \in \mathbb{R}^{d_l}$ contain the bias terms and activations in layer H_l.

The hidden layers of a feedforward network can be seen as a feature transformation function. In the first hidden layer, the input x is transformed into a d_1 dimensional feature vector. In H_2, this vector is transformed into a d_2 dimensional vector an so on. Therefore, for layer l we have

$$a^l = \mathcal{A}_l \left(w^l a^{l-1} + b^l \right).$$

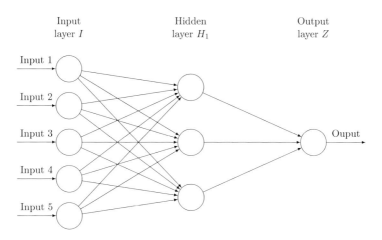

Figure 6.2: Illustration of an artificial neural network.

Finally, the output layer Z classifies this d_{L-1} dimensional feature vector. The output of the network can be formulated by

$$f(x, W) = a^L = \mathcal{A}_L \left(\cdots \left(\mathcal{A}_2 \left(\mathcal{A}_1 \left(w^1 x + b^1 \right) w^2 + b^2 \right) \cdots \right) w^L + b^L \right),$$

where we set $W := \left\{ w^1, \ldots, w^L, b^1, \ldots b^L \right\}$. The feature transformation of a feedforward network does not have to be designed by hand but can be learned instead. We only have to define the number of hidden layers, number of neurons in each layer and the type of activation functions. These inputs are called *hyperparameters*. Given these parameters, feature transformation as well as classification can be trained. The objective during training is to minimize the error (or loss) of the network output. We consider a training data set $\mathbb{D} = \{(x_1, y_1), \ldots (x_n, y_n)\}$ with inputs $x_k \in \mathbb{R}^d$ and labels $y_k \in \mathbb{R}^m$, where $m \in \mathbb{N}$ is the number of output neurons. Given this notation, we can formulate a loss function which measures the error between the outputs of the network $f(x_k, W) := z_k = (z_{k,1}, \ldots, z_{k,m})$ and the true labels $y_k = (y_{k,1}, \ldots, y_{k,m})$. For example, we can define the square loss function

$$\mathcal{L}(W) := \frac{1}{2} \sum_{k=1}^{n} \sum_{o=1}^{m} (z_{k,o} - y_{k,o})^2.$$

Next, the procedure is to minimize such a loss function by the *stochastic gradient descent* algorithm, which is a so-called *batch* procedure. That means, not all training samples are taken into account for the calculation of the loss but only a randomly chosen subset of a predefined batch size. After initialization, W is changed iteratively proportional to the gradient of loss $\nabla \mathcal{L}$, i.e.

$$\Delta W = -\bar{\eta} \nabla \mathcal{L}(W).$$

The constant $\bar{\eta} \in \mathbb{R}$ is called *learning rate*. As described in [2], a direct computation of this gradient is not tractable in practice due to the usual huge amount of layers and neurons. To

Figure 6.3:

overcome this issue, an algorithm called *backpropagation* [121] can be used to compute this gradient of loss. For each iteration, the new weights and bias terms are calculated as follows

$$\Delta w^l = -\bar{\eta} \delta^l a^{l-1},$$
$$\Delta b^l = -\bar{\eta} \delta^l$$

with

$$\delta_j^L = \mathcal{A}_L' \left(\sum_{i=1}^{d_{L-1}} w_{i,j}^L a_j^{L-1} + b^L \right) \left(y_j - a_j^L \right),$$

$$\delta_j^l = \mathcal{A}_l' \left(\sum_{i=1}^{d_{l-1}} w_{i,j}^l a_j^{l-1} + b^l \right) \left(\left(w^{l+1} \right)^T \delta^{l+1} \right)$$

for all neurons j in the corresponding layer.

As stated before, there are different types of activation functions which can be adapted to the problem to be solved. According to [2], nonlinear activation functions in at least one neuron of a feedforward network is required to be able to learn a nonlinear function. With linear activation functions in all neurons, a feedforward network can only learn linear functions. A second important property of an activation function is its differentiability since learning is usually done by a gradient descend method.

As described in [2], a popular activation function is the *sigmoid* function $\mathcal{A}_{sig} \colon \mathbb{R} \to [0,1]$, which is given by

$$\mathcal{A}_{sig}(x) := \frac{1}{1 + e^{-x}}.$$

Its derivative can be expressed by

$$\mathcal{A}_{sig}'(x) = \mathcal{A}_{sig}(x) \left(1 - \mathcal{A}_{sig}(x) \right).$$

Figure 6.4a shows the plots for these two functions.

The sigmoid function fulfills the requirement of differentiability. But the drawback is that its gradients become very small for $|x| \to \infty$. During training with the backpropagation method, this causes a problem called *vanishing gradients problem*. During backpropagation, the gradient of the loss function with respect to the current node is calculated by multiplying the gradient of the activation function with its children. Thus, if the input for the activation function is far from zero, then the gradient of loss with respect to the current node will be very small. In case we have a network with many sigmoid functions as activations, the gradient vanishes in the first layers. This leads to very small weight changes and a stop in learning. Therefore, the sigmoid activation function can mainly be used in shallow networks, i.e. networks with few hidden layers. It is not suitable for training networks with many hidden layers, also called deep networks.

Another possibility is the *rectified linear unit (ReLU)* activation function, which is defined by

$$\mathcal{A}_{relu}(x) := \max(0, x).$$

The derivative of the ReLU function is given by

$$\mathcal{A}_{relu}'(x) = \begin{cases} 0 & \text{for } x < 0 \\ 1 & \text{for } x \geq 0 \end{cases}.$$

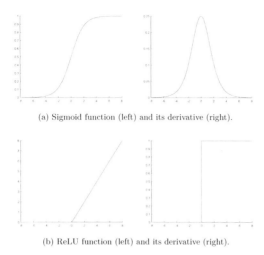

(a) Sigmoid function (left) and its derivative (right).

(b) ReLU function (left) and its derivative (right).

Figure 6.4: Plots of commonly used activation functions.

Their plots are shown in Figure 6.4b. According to [2], this nonlinear function is a good choice for deep networks since it always delivers a strong gradient and thus avoids the vanishing gradient problem. There are several types of activation functions building up on the ReLU function. Since they do not provide further insights into the basic function of a neural network, we refrain from their description.

6.1.2   Convolutional Neural Networks

As pointed out in [2], the structure of fully-connected feedforward networks is not suitable for an application on images. Each image pixel needs to be connected to a neuron in the input layer I. Next, each neuron in the input layer is connected with each neuron in the first hidden layer. For a $M \times N$ sized grayscale image $f \in \mathbb{R}^{M \times N}$ and d_1 neurons in the first hidden layer H_1 we have a total number $M \cdot N \cdot d_1$ connections between I and H_1. Even for images with moderate resolution and shallow networks this would result in an impracticable huge amount of neurons involved in the network. For example, considering a 16×16 grayscale image and $7,200$ neurons in H_1, we already have $16^2 \cdot 7,200 = 7,372,800$ parameters for connecting I with H_1. For this reason, *Convolutional Neural Networks (CNNs)* have been developed with the aim to reduce the number of network parameters.

The Way to Convolution

The first strategy to reduce the number of parameters is to consider the geometry of pixels in an image. For a given image pixel (m, n), we have a stronger correlation to its near than to its far neighbors. Therefore, we only use the information of pixel (m, n) and the pixels in a region of predefined size around it to extract information. Specifically, we first rearrange the d_l neurons in a hidden layer H_l into a number of B_l blocks of size $M_l' \times N_l'$. Next, we connect each neuron in each block to a $K_l \times K_l$ sized image region. This region is called *receptive field* of the corresponding neuron. In each block, the neurons in it extract information from each $K_l \times K_l$ image patch. Thus, the neurons in each block cover the whole image. In our above example, with $B_1 = 50$, M_1', $N_1' = 12$ and $K_1 = 5$, we can reduce the number of network parameters to $5^2 \cdot 50 \cdot 12^2 = 180,000$.

As a second strategy for further reduction, we assume that all weights in one block share the same weights. In our example, this leads to just $5^2 \cdot 50 = 1,250$ weights, i.e. connections from I to H_1. This is just 0.017% of the original number of parameters in a fully-connected network. For a further description of this *weight sharing* technique, we denote the output matrix of block b in layer H_1 by n^b. Then, we have for all neurons $(p,q) \in \{0, \dots M_1' - 1\} \times \{0, \dots N_1' - 1\}$

$$n^b(p,q) = \mathcal{A}\left(\sum_{j=0}^{K_1-1}\sum_{k=0}^{K_1-1} f(p+j, q+k) w_{j,k}^b\right), \tag{6.1}$$

where $w_{j,k}^b$ denotes the weight $(j,k) \in \{0, \dots, K_1 - 1\}^2$ in block b of layer H_1. We can consider (6.1) as the result of the *discrete convolution* of the input image f with the $K_1 \times K_1$ sized filter w followed by an element-wise application of the activation function $\mathcal{A}$. For the output of the layer, convolution and activation function application is performed for each filter. The output is then given by several image transformations, also called *feature maps*. More precisely, for our $M \times N$ grayscale image, we get B_1 feature maps of size $(M - K_1 + 1) \times (N - K_1 + 1)$. Of course, the described procedure can not solely be applied in H_1 to the input image as in (6.1) but also in subsequent layers to feature maps. Any layer performing convolution on either the input image or on feature maps is called *convolutional layer*.

In CNNs, a convolution of a filter with a multi-channel input, such as a RGB image or a feature map collection, results in single-channel output. Let $X \in \mathbb{R}^{M \times N \times C}$ be multi-channel input for convolution with C channels. To obtain a single-channel output of the convolution with a filter g, we need the filter to be three-dimensional also with C channels, i.e. $g \in \mathbb{R}^{M' \times N' \times C}$. Then we have $X * g \in \mathbb{R}^{M-M'+1 \times N-N'+1 \times 1}$.

We now assume that we have a multi-channel image $f \in \mathbb{R}^{M \times N \times C}$ as input for the first convolutional layer with B_1 filters of size $M_1 \times N_1 \times C$. Thus, we get a feature map collection $f_1 \in \mathbb{R}^{M-M_1+1 \times N-N_1+1 \times B_1}$ after the first convolutional layer. The B_2 filters of a consecutive, second convolutional layer must then have dimensions $M_2 \times N_2 \times B_1$. Consequently, we get B_2 feature maps after the second convolutional layer.

CNNs incorporate the possibility to define a *stride of convolution* $s_{conv} \in \mathbb{N}$. A stride $s_{conv} > 1$ means that the convolution is not computed for all pixels. For convolving $X \in \mathbb{R}^{M \times N}$ with filter $g \in \mathbb{R}^{M' \times N'}$, we consider the formula

$$(X * g)(m,n) = \sum_{i=0}^{M'-1}\sum_{j=0}^{N'-1} X(m+1, n+j) g(i,j),$$
$$m = 0, s_c, 2s_{conv}, \dots, M-1, n = 0, s_{conv}, 2s_{conv}, \dots, N-1.$$

Consequently, the result has dimensions $\frac{M-M'}{s_{conv}} + 1 \times \frac{N-N'}{s_{conv}} + 1$.

Incorporation of Pooling

Similarly to hand-crafted feature detectors, CNNs incorporate an approach for feature pooling. For this task, pooling layers are set up in the network architecture. The main reason for pooling is a reduction of the dimensions, i.e. *downsampling*, of the feature maps. The downsampling

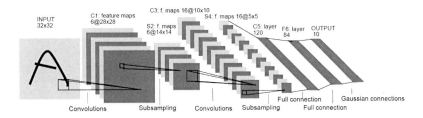

Figure 6.5: LeNet-5 architecture [82].

factor is also called *pooling stride* and we denote it with $s_{pool} \in \mathbb{N}$. To not lose too much data during downsampling, a method called *max pooling* is commonly used. The pooling input feature map is partitioned into $d_{pool} \times d_{pool}$ sized patches every s_{pool} pixels row- and column-wise. For each patch we then have a pixel in the pooling output feature map showing the maximum value in the corresponding patch. If $s_{pool} < d_{pool}$, the pooling patches will overlap with their neighbors. The pooling is applied to each feature map separately. Thus, the number of output feature maps equals the number of input feature maps.

As an alternative to max pooling we have *average pooling*. As the name indicates not the maximum value of a feature map patch is set as output but its average. According to Scherer et al. [110], this pooling procedure usually provides inferior results compared to max pooling.

CNN Architecture

We now will put together the components of a CNN which we described above. Usually, a CNN is composed of several convolutional layers with pooling layers in between. Directly before the output layer, we have few fully-connected layers.

According to [2], the first layers usually have a small number of feature maps and the number increases with the depth of the network. Furthermore, convolutional filters of sizes 3×3, 5×5, 7×7 and 11×11 are commonly used. Activation functions are usually placed directly after a convolution layer. In practice, the usage of ReLU type activation function has been established.

We will further describe the interaction of the network layers by the example of one of the first popular implementations of CNNs, called *LeNet-5*. This network has been developed by LeCun et al. [82] for the classification of handwritten digits. The architecture of LeNet-5 is shown in Figure 6.5. It consists of two convolutional layers, two pooling layers, two fully-connected layers and an output layer. The input for this network is a grayscale image of size 32×32. The first convolution layer C1 consists of 6 filters with size $5 \times 5 \times 1$. The size of the resulting 6 feature maps is reduced to 14×14 by pooling layer S2 with $d_{pool} = s_{pool} = 2$. The second convolutional layer C3 has 16 filters of size $5 \times 5 \times 6$. Consequently, its output is of dimensions $10 \times 10 \times 16$. The pooling layer S4 then generates 16 feature maps of size 5×5. Subsequently, we have a fully-connected layer C5 with 120 neurons, which are all connected to all neurons in S4. The layer C5 can also be regarded as a convolutional layer with 120 filters of size 5×5. Following, another fully-connected layer F6 with 84 neurons is included. The classification is performed by a radial basis function which is using 84-dimensional vectors as inputs. Finally, the output layer is a 10-dimensional vector since 10 different digits shall be classified.

In the next years after the development of LeNet-5, the number of involved layers increased. In 2012, the popular CNN called *AlexNet*, developed by Krizhevsky et al. [72], involves 5

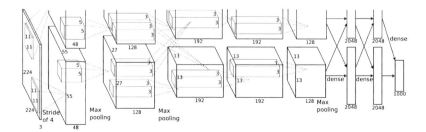

Figure 6.6: AlexNet architecture [72].

convolution and 3 pooling layers. LeNet-5 only uses 3 convolution and 2 pooling layers. With a training on the huge ImageNet data set [24], Krizhevsky et al. [72] were able to win the image classification task of the *ImageNet Large-Scale Visual Recognition Challenge (ILSVRC)* in 2012. The data set contains around 1.2 millions training images, $50,000$ validation images and $150,000$ test images of 1000 classes of natural objects. The task of the ILSVRC classification competition is to classify color images into the predefined 1000 distinct classes. Consequently, the output of AlexNet is a 1000-dimensional vector. In this chapter, we pay special attention to AlexNet since it uses relatively big filters of size 11×11 in the first layer. This size is suitable to set up local precision shearlet filters in contrast to smaller sizes often used in other CNN architectures. Figure 6.6 shows the architecture of AlexNet. Two GPUs have been used in the original implementation. The top layer part in the figure is carried out by one GPU, the lower part by the other.

In the following years of the ImageNet challenge, the number of layers continued to increase [63, 113, 115], reaching over 150 layers. One of these networks, namely the *VGG16* [113], is used in the current top performing algorithms for pedestrian detection [9, 32, 127]. Therefore, we also consider this network for the integration of shearlets. The VGG16 network uses 13 convolutional layers, 5 pooling layers and 3 fully-connected layers. Therefore, it has 16 layers containing weights. One of the main characteristics of this network is that is uses 3×3 filters in all convolution layers. Since this size is too small to define multiscale shearlet filters, we need to adapt this architecture for our shearlet integration. We will describe our adaptation in Section 6.2.2.

6.1.3   Pedestrian Detection Algorithms using CNNs

As described by Benenson et al. [8], the first approaches using CNNs for the task of pedestrian detection did not yield an improvement compared to hand-crafted feature detectors, although CNNs already dominated classification challenges at that time. A reason behind this observation is that object detection is much more complex than its classification. Objects shall not only be classified but also localized without detailed information concerning size and number of objects in the image.

Only by the development of the *R-CNN* approach of Girshick et al. [42], this problem could be resolved. R-CNN is the abbreviation for *Regions with CNN features*. This approach utilizes *Region Proposals*. That means, *Regions of Interest (RoI)* are extracted first by an external algorithm. After these regions are scaled to a specific size, features for each RoI are extracted

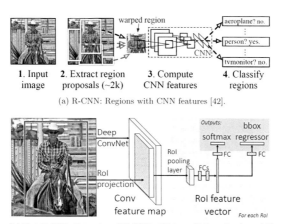

1. Input   2. Extract region   3. Compute   4. Classify
image   proposals (~2k)   CNN features   regions

(a) R-CNN: Regions with CNN features [42].

(b) Fast R-CNN: Extension of R-CNN utilizing a RoI pooling layer [41].

Figure 6.7: illustration of the first successful detection algorithms utilizing CNNs.

by a CNN. These features are then learned by an SVM classifier per class for the classification of each RoI. Figure 6.7a illustrates the procedure of R-CNN.

Main drawbacks of R-CNN are the runtime and the expensive training. The R-CNN procedure is relatively slow, since CNN features have to be computed for each RoI separately. For deep networks, this approach results in time consuming computations that are performed multiple times in case RoIs are overlapping. Furthermore, the training has to be performed in several steps. First, a CNN fine-tuning is performed followed by a training of SVM according to the CNN features. Finally, a bounding box regressor is trained to reduce localization errors.

To resolve these drawbacks, Girshick [41] developed the *Fast R-CNN* algorithm by the introduction of *RoI pooling layers*. Figure 6.7b illustrates the Fast R-CNN approach. An important insight for this approach is that the weights of convolutional layers are independent of the input size. Only the fully-connected layers require a fixed input size. The purpose of the RoI-Pooling is to project the features provided by the last convolutional layer onto a fixed sized feature vector. Like the R-CNN algorithm, Fast R-CNN receives an image and RoIs as input. Then, it computes a feature map of the image by a convolutional network and projects the RoIs onto this feature map. For that, the RoI pooling layer uses max-pooling to convert the feature map of each RoI to a fixed size. With the aid of subsequent fully-connected layers, a feature vector for each RoI is provided. The output of the algorithm contains two components. First, it determines the object class with the highest classification certainty for each region and second the bounding box for each object.

Still, the Fast R-CNN approach has the drawback that it requires RoIs computed by an external algorithm. Therefore, the detection quality of Fast R-CNN is depending on the quality of the precomputed Region Proposals. To resolve this dependency, Ren et al. [109] developed the *Faster R-CNN* algorithm. It is an extension of Fast R-CNN by inserting a so-called *Region Proposal Network (RPN)* after the last convolutional layer. The task of this network is to determine RoIs, which are then given the Fast R-CNN algorithm as input. The general architecture of Faster R-CNN is shown in Figure 6.8.

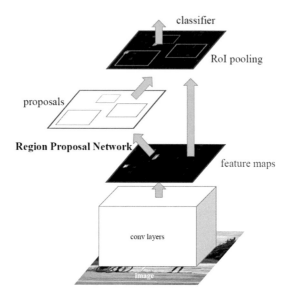

Figure 6.8: Architecture of Faster R-CNN [109].

The RPN operates as a sliding window detector, i.e. it is slid over the feature map while an evaluation takes place at each location of it. This procedure is carried out by a convolutional layer with filters of small size, e.g. 3×3. In that way, for each position of the slid window, another feature is computed, which is given to two fully-connected layers as input. These layers determine if an object is present in the corresponding region and as the case may be, the bounding box of it. Since there can be multiple objects within one region, a fixed number of so-called *anchors* is considered for each region. These anchors represent different scales and aspect ratios of an object. In [109], three scales are used with three aspect ratios respectively, i.e. nine anchors in total. Finally, the outputs of the fully-connected layers are determined in relation to the anchors. Figure 6.9 shows the structure of the Region Proposal Network assuming that the last layer contains 256 filters.

Faster R-CNN provides very good results in relatively short runtime for general object detection tasks, see [109]. However, Zhang et al. [127] show that this approach has an issue with the specific task of pedestrian detection measured on the Caltech data set. Interestingly, the results of the RPN part are quite good but the detection rates decline after utilization of the Fast R-CNN algorithm. The authors reason this finding with the small pixel size of pedestrians in the Caltech data set, whereas the filters of the last convolutional layer operate on a large image section. Due to this mismatch, useful information may get lost in the RoI pooling layer. Furthermore, Zhang et al. [127] argue that general object detection and pedestrian detection have different difficulties. In general object detection, the main root cause for false predictions is the presence of multiple object classes. In contrast, pedestrian detection mainly struggles with hard background instances similar to pedestrians, such as streetlights.

Derived from their findings, Zhang et al. [127] developed the *RPN+BF* algorithm. Since RPN delivers quite good results, this procedure is used to generate bounding boxes and corresponding

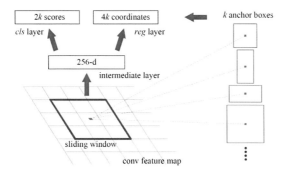

Figure 6.9: Structure of the Region Proposal Network [109].

detection certainties. In combination with the features computed by the RPN, these outputs are given a Boosted Forrest of [38] as input to provide the final results. Specific anchors reflecting the properties of pedestrians are used for the RPN part. Nine different scales are used but only one aspect ratio with value 0.41, which is the mean aspect ratio of pedestrians in the Caltech data set. Similar to the AdaBoost classifier we used in Chapter 5, the Boosted Forrest classifier is trained in multiple steps with bootstrapping. To resolve the issues caused by small sized pedestrians, not only the features of the last convolutional layer are used but also features of previous layers. The RPN+BF algorithm achieves a log-average miss rate of 9.6%, which made it the best known approach at the date of its publication. In comparison, Faster R-CNN only achieves 20.2%.

A subsequent approach was introduced by Du et al. [32]. This algorithm is called Fused DNN, it is illustrated in Figure 6.10. Similar as in the RPN+BF approach, the first step is the generation of pedestrian candidates. In this case, these candidates are provided by the *Single shot multibox detector (SSD)* [89]. The SSD uses the VGG16 architecture as a basis but adds further convolutional layers at the end of the network. In total seven layers generate the pedestrian candidates by which different image scales are considered. The threshold for the detection certainty is set to a very small value of 1% such that as many pedestrians as possible are detected. However, this produces many false positives.

The generated candidates are then given as input to one or more classification networks with the aim to scale their confidence scores. Let p_m be the score that classification network m provides for a candidate. The SSD score S_{SSD} is then multiplied by a factor

$$a_m := \max\left(\frac{p_m}{a}, b\right),$$

where the required constants are set as $a = 0.7$ and $b = 0.1$ by cross validation, see [32]. Consequently, the confidence gets increased if the network classifies the candidate as a pedestrian with a score $p_m > a$. Otherwise it is scaled by a factor $a_m < 1$ but not less than b in order to prevent any classification network from dominating. The score computed by the algorithm can be expressed by

$$S_{FDNN} = S_{SSD} \prod_{m=1}^{M} a_m,$$

where M is the number of classification networks used. In [32], the utilization of the two networks *GoogLeNet* [115] and *ResNet-50* [63] yields a log-average miss rate of 8.65% for the

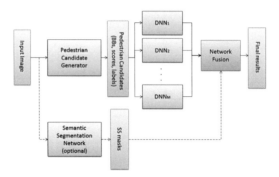

Figure 6.10: Architecture of Fused DNN [32].

Caltech benchmark. In addition, Du et al. [32] utilize the *semantic segmentation* network of [126] for a parallel classification in order to further improve the detection rates. This network classifies each pixel to the two classes "pedestrian" and "no pedestrian". The score S_{FDNN} is preserved in case the pixels belonging to a pedestrian account for at least 20% of the bounding box. Otherwise, the score is reduced with respect to size of overlapping area. The combined score is expressed by

$$S_{all} = \begin{cases} S_{FDNN}, & \text{if } \frac{A_m}{A_b} > 0.2 \\ S_{FDNN} \max\left(\frac{A_m}{A_b} a_{ss}, b_{ss}\right), & \text{otherwise,} \end{cases}$$

where A_b is the area of the bounding box and A_m the area within A_b belonging to a pedestrian in the semantic segmentation mask. Again, the required constants are set by cross validation with $a_{ss} = 4$ and $b_{ss} = 0.35$. With the utilization of the semantic segmentation network, the log-average miss rate improves to 8.18%. Not surprisingly, the runtime increases significantly. Du et al. [32] report an increase from 0.16s per frame to 2.48s per frame with the usage of a Titan X GPU.

Finally, we present the approach of Brazil et al. [9], named *simultaneous detection and segmentation R-CNN (SDS-RCNN)*. This procedure combines insights from RPN+BF and Fused DNN and achieves currently the best results in the Caltech benchmark. The architecture of SDS-RCNN is illustrated in Figure 6.11a. The procedure consists of two stages. First, an RPN is used for the generation of pedestrian candidates and corresponding scores. As basis for the computation of feature maps, Brazil et al. [9] use layers of the VGG16 model pretrained on ImageNet. The generated candidates are then given to a classification network similar as in [32]. This network also consists of VGG16 layers including the first two fully-connected layers. Again similar to [32], the candidate scores are adjusted by the classification network. Let p_i^1, $i \in \{0, 1\}$, be the RPN score for class i and p_i^2 the score of the classification network accordingly. The final score is computed by

$$p = \frac{e^{p_1^1 + p_1^2}}{e^{p_1^1 + p_1^2} + e^{p_0^1 + p_0^2}}.$$

Consequently, an exceedingly confident score is obtained in case both networks yield the same result. Otherwise, the result corresponds to the value of the more confident network. Therefore, it is desirable that at least one network is confident with its classification in case of differing results provided by the two networks.

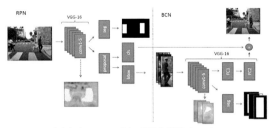

(a) Architecture of the SDS-RCNN algorithm.

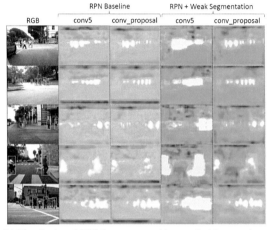

(b) Visualization of RPN feature maps without and with semantic segmentation.

Figure 6.11: Illustration of the key features of the SDS-RCNN procedure [9].

In order to use the fully-connected layers of the VGG16 model for the classification network, the candidates have to be scaled to a fixed size. The original VGG16 network requires an input image of size 224×224. Since the most pedestrians in the Caltech data set have a height between 30 and 80 pixels, the input image size is adjusted to 112×112 and the last pooling layer is discarded. With the insight of [32] that the incorporation of semantic segmentation can improve the detection rates, Brazil et al. [9] insert another layer in both stages of SDS-RCNN. The so-called *segmentation infusion layer* outputs two masks in order to *illuminate* pedestrians in the feature maps preceding the classification layers. The infusion layer is placed after the fifth convolutional layer, where the input data has been down-sampled significantly. Therefore, Brazil et al. [9] find that it is sufficient to train the infusion layer with box-based annotations instead of pixel-wise ones. Figure 6.11b visualizes the influence of the segmentation infusion layer on the RPN feature maps. Since the segmentation is performed on already computed features, its incorporation does not effect the runtime as much as for Fused DNN [32].

The training of SDS-RCNN proceeds in two phases. First, only the RPN is trained, whereas the weights of the first four convolutional layers are set as fixed. Subsequently, the classification network is trained with pedestrian candidates generated by the RPN. Also the first four layers of the classification network are set as fixed. As a result, two different feature maps are obtained by the networks. The SDS-RCNN algorithm achieves a log-average miss rate of 7.36%, which is currently the best result in the Caltech Pedestrian Detection Benchmark. The Matlab code of it is available online under https://github.com/garrickbrazil/SDS-RCNN.

6.2 Integration of Shearlets in CNNs

As we have seen in the previous chapter, currently all best performing pedestrian detection algorithms in the Caltech benchmark are based on CNNs. However, it may be possible to achieve even better results if one incorporates "traditional" object detection approaches and theoretical frameworks. In Chapter 5, we showed that shearlets provide the framework to achieve the best performing hand-crafted features for pedestrian detection. Therefore, we integrate shearlets in a CNN and analyze in which way we can benefit from it.

In the following, we first have a look at publications related to our approach. Here, we find insights about the functionality of CNNs justifying our basic idea. Subsequently, we derive a concept for the integration of shearlets in neural networks. Finally, we present the implementation framework we use in our work. Especially, we describe the deep learning framework Caffe and our strategy to incorporate shearlets in it.

6.2.1 Related Work

In [116], Szegedy et al. find an instability of deep neural networks against certain perturbations of its input. By small variations of test images with the aim to maximize the prediction error of the network, the authors create *adversarial images*. These perturbations are imperceptible to the eye but they cause the network to misclassify the image. Furthermore, the applied perturbations are robust in a sense that neural networks with different learning parameters and training data still are subject to the same misclassification. Thus, a theoretical treatment of neural networks should especially investigate the stability under various transformations such as scalings, translations, deformations.

A first approach to provide such a stability is made by Bruna and Mallat [10]. Here, the wavelet *scattering transform* introduced in [92] is implemented by a convolutional network. Bruna and Mallat set up two-dimensional wavelets by dilating a band-pass filter ψ by 2^j for $j \in \mathbb{Z}$ and by rotating by $r_\theta \in \mathcal{R}_\theta$, where $\mathcal{R}_\theta$ is a set of rotations of angles $\theta = 2k\pi/K$ for $0 \le k < K \in \mathbb{N}$, i.e.

$$\psi_{j,\theta}(x) := 2^{-2j} \psi\left(2^{-j} r_\theta x\right). \tag{6.2}$$

The corresponding wavelet transform of a signal $f \in L^2(\mathbb{R}^2)$ with respect to j and r_θ is computed by $f * \psi_{j,\theta}$. For a sequence $p = ((j_1, \theta_1), (j_2, \theta_2), \dots, (j_l, \theta_l))$ also called *path* of length $l \in \mathbb{N}$, the scattering transform is given by

$$\widetilde{S}_p(f) := \mu_p^{-1} \int_{\mathbb{R}^2} \mathcal{U}_p f(x)\, \mathrm{d}x,$$

with

$$\mathcal{U}_p f := |||f * \psi_{j_1,\theta_1}| * \psi_{j_2,\theta_2}| \cdots * \psi_{j_l,\theta_l}|$$

and the Dirac response

$$\mu_p := \int_{\mathbb{R}^2} \mathcal{U}_p \delta(x)\, \mathrm{d}x.$$

The scattering transform is a translation invariant representation which is Lipschitz-continuous to deformations. For setting up a network, a windowed scattering transformed in the neighborhood of $x \in \mathbb{R}^2$ is used which is defined by

$$
\begin{aligned}
\mathcal{S}_p f(x) &:= \mathcal{U}_p f(x) * \phi_{2^J}(x) = \int_{\mathbb{R}^2} \mathcal{U}_p f(y) \phi_{2^J}(x-y) \, dy \\
&= \left| \left| \left| f * \psi_{j_1,\theta_1} \right| * \psi_{j_2,\theta_2} \right| \cdots * \psi_{j_l,\theta_l} \right| * \phi_{2^J}(x),
\end{aligned}
$$

with a scaled low-pass filter $\phi_{2^J}(x) = 2^{-2J} \phi(2^{-J} x)$ and a predefined scale 2^J. The layer l of the scattering convolution network is defined by the propagated signals $(\mathcal{U}_p f)_{p \in \mathcal{P}_l}$, where $\mathcal{P}_l$ is the set of all paths $p := ((j_1, \theta_1), \ldots, (j_l, \theta_l))$ of length l. Correspondingly, the filters are not learned but given by predefined wavelets. Bruna and Mallat [10] employ this network to the task of classification of handwritten digits and achieve state-of-the-art results.

Oyallon et al. [104] utilize a network with wavelet scattering to the task of object classification. This network is extended by Oyallon and Mallat [103] by the application of a second order wavelet transform. Given an image $f \in L^2(\mathbb{R}^2)$, the output of the network with J layers, denoted by f_J, is computed by two wavelet transforms. At a given depth j_1 in the network, the absolute value of the spatial wavelet transform at scale j_1 is computed. The output $f_{j_1}^1$ can be expressed by

$$
f_{j_1}^1(x, \theta) := \left| f * \psi_{j_1, \theta} \left(2^{j_1 - 1} x \right) \right|.
$$

In order to obtain a representation which is stable to rotations and to deformations along rotations, a second wavelet transform is computed along the angle parameter θ. This transform is called *roto-translation wavelet transform*. It is computed by convolutions with the three dimensional wavelet

$$
\psi_{j, \beta, \tilde{j}}(x, \theta) := \psi_{j, \beta}(x) \overline{\psi_{\tilde{j}}}(\theta),
$$

with a spatial wavelet $\psi_{j, \beta}(x)$ as in (6.2) of scale $j > j_1$ and an angular wavelet $\overline{\psi_{\tilde{j}}}(\theta)$ of scale $2^{\tilde{j}}$ for $1 \leq \tilde{j} \leq \tilde{J} < \log_2 K$. The output f_j^2 for scale $j > j_1$ is given by

$$
f_j^2(x, \theta) := \left| f_{j_1}^1 * \psi_{j, \beta, \tilde{j}} \left(2^{-j-1} x, 2^{-\tilde{j}-1} \theta \right) \right|.
$$

For the final output of the network at layer J, the image as well as the first and the second order coefficients are averaged at scale 2^J, i.e.

$$
f_J := \left\{ f * \phi_{2^J}, f_j^1 * \phi_{2^J}, f_j^2 * \phi_{2^J} \right\}_{1 \leq j \leq J},
$$

with the scaled low-pass filter ϕ_{2^J}. According to Oyallon and Mallat [103], the instability against perturbations found by Szegedy et al. [116] can be avoided by the usage of the roto-translation wavelet transform. The roto-translation scattering networks achieve comparable results to unsupervised deep learning approaches. However, supervised learning procedures score significantly better results. An increase of the number of wavelet layers does not yield a classification improvement. To further improve the classification results, Oyallon et al. [102] introduce *Deep Hybrid Networks*. The first layers of these networks are initialized and fixed by roto-translations followed by a common CNN architecture. This way, comparable results to standard CNN approaches can be achieved while less layers have to be learned.

All approaches presented above are based on the work of Mallat [92]. They are restricted to the application of wavelets in convolutional networks. Therefore, Wiatkowski and Bölcskei [122, 123] extend their theoretical analysis to more general cases. More precisely, the layer l of a network is build up on a *module* consisting of a semi-discrete frame and Lipschitz-continuous operators.

Based on atoms $g_{\lambda_l} \in L^1(\mathbb{R}^2) \cap L^2(\mathbb{R}^2)$, a semi-discrete frame for $L^2(\mathbb{R}^2)$ is a collection $\Psi_l := \overline{\{T_b I g_{\lambda_l}\}_{b \in \mathbb{R}^2, \lambda_l \in \Lambda_l}}$, with translation $(T_t f)(x) = f(x - t)$, $t \in \mathbb{R}^2$, involution $(If)(x) := \overline{f(-x)}$ and a countable index set Λ_l, if there exist $0 < A \leq B < \infty$ such that

$$A \|f\|_2^2 \leq \sum_{\lambda_l \in \Lambda_l} \int_{\mathbb{R}^2} |\langle f, T_b I g_{\lambda_l} \rangle|^2 \, \mathrm{d}b = \sum_{\lambda_l \in \Lambda_l} \|f * g_{\lambda_l}\|_2^2 \leq B \|f\|_2^2 ,$$

for all $f \in L^2(\mathbb{R}^2)$. The following definitions and statements are abstracted from [123].

Definition 6.1. For $l \in \mathbb{N}$, let Ψ_l be a semi-discrete frame for $L^2(\mathbb{R}^2)$ and let $M_l \colon L^2(\mathbb{R}^2) \to L^2(\mathbb{R}^2)$ and $P_l \colon L^2(\mathbb{R}^2) \to L^2(\mathbb{R}^2)$ be Lipschitz-continuous operators with $M_l f = 0$ and $P_l f = 0$, respectively. Then, the sequence of triplets

$$\Omega := ((\Psi_l, M_l, P_l))_{l \in \mathbb{N}}$$

is called a *module-sequence*. In each frame of the module-sequence, one of the atoms is designated as the *output-generating atom* $\chi_{l-1} := g_{\lambda_l^*}$, $\lambda_l^* \in \Lambda_l$, of the network layer $l - 1$.

Based on such a module-sequence, the operator of a network layer and the *feature extractor* of a network are defined as follows. The corresponding network architecture is illustrated in Figure 6.12.

Definition 6.2. Let $\Omega = ((\Psi_l, M_l, P_l))_{l \in \mathbb{N}}$ be a module-sequence, let $\{g_{\lambda_l}\}_{\lambda_l \in \Lambda_l}$ be the atoms of the frame Ψ_l and let S_l be the pooling factor associated with the network layer l. Define the operator $\widetilde{U}_l$ associated with layer $l \in \mathbb{N}$ of the network as $\widetilde{U}_l \colon \Lambda_l \times L^2(\mathbb{R}^2) \to L^2(\mathbb{R}^2)$,

$$\widetilde{U}_l (\lambda_l, f) := \widetilde{U}_l [\lambda_l] f := S_l P_l (M_l (f * g_l)) (S_l \cdot) .$$

Define the set $\Lambda_1^l := \Lambda_1 \times \Lambda_2 \times \cdots \times \Lambda_l$ as well as $\Lambda_1^0 := \{\emptyset\}$ and $\widetilde{U}_0 [\emptyset] f := f$ for all $f \in L^2(\mathbb{R}^2)$. Furthermore, the operator is extended for $q \in \Lambda_1^l$ to

$$\begin{aligned} \widetilde{U} [q] f &= \widetilde{U} [(\lambda_1, \lambda_2, \ldots, \lambda_l)] f \\ &:= \widetilde{U}_l [\lambda_l] \cdots \widetilde{U}_2 [\lambda_2] \widetilde{U}_1 [\lambda_1] f , \end{aligned}$$

with $\widetilde{U} [\emptyset] f := f$.

Then, the *feature extractor* Φ_Ω based on Ω maps $f \in L^2(\mathbb{R}^2)$ to its feature vector

$$\Phi_\Omega (f) := \bigcup_{l=0}^{\infty} \Phi_\Omega^l (f) ,$$

where $\Phi_\Omega^l (f) := \left\{ \left(\widetilde{U} [q] f \right) * \chi_l \right\}_{q \in \Lambda_1^l}$, with $\chi_{l-1} := g_{\lambda_l}$ for all $l \in \mathbb{N}$.

According to Wiatkowski and Bölcskei [123], the feature extractor Φ_Ω is translation invariant in a way that it becomes more translation invariant as the network depth increases. Furthermore, the authors provide a bound on the sensitivity of the feature extractor with respect to time-frequency deformations defined by

$$(F_{\tau, \omega} f) (x) := e^{2\pi i \omega(x)} f (x - \tau (x)) ,$$

with $\omega \in C(\mathbb{R}^2, \mathbb{R})$ and $\tau \in C^1(\mathbb{R}^2, \mathbb{R}^2)$. We summarize these results in the following theorem.

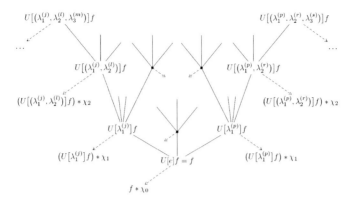

Figure 6.12: Network architecture of a deep CNN feature extractor Φ_Ω [123]. The index $\lambda_l^{(k)}$ is associated with the k-th atom $g_{\lambda_l^{(k)}}$ of frame Ψ_l in network layer l.

Theorem 6.3. *Let* $\Omega = ((\Psi_l, M_l, P_l))_{l \in \mathbb{N}}$ *be a module-sequence with frame upper bounds* $B_l > 0$ *and Lipschitz constants* L_l, $R_l > 0$ *of the operators* M_l *and* P_l *such that*

$$\max\left(B_l, B_l L_l^2 R_l^2\right) \leq 1, \quad \text{for all } l \in \mathbb{N}.$$

i. *Let* $S_l \geq 1$, $l \in \mathbb{N}$, *be the pooling factor of the network layer* l *and assume that the operators* M_l *and* P_l *commute with the translation operator* T_t, *i.e.*

$$M_l T_t f = T_t M_l f, \quad P_l T_t f = T_t P_l f,$$

for all $f \in L^2(\mathbb{R}^2)$, $t \in \mathbb{R}^2$ *and* $l \in \mathbb{N}$. *The features* $\Phi_\Omega^l(f)$ *generated in network layer* l *satisfy*

$$\Phi_\Omega^l(T_t f) = T_{t/(S_1 \cdots S_l)} \Phi_\Omega^l(f),$$

for all $f \in L^2(\mathbb{R}^2)$, $t \in \mathbb{R}^2$ *and* $l \in \mathbb{N}$, *where* $T_t \Phi_\Omega^l(f)$ *refers to the element-wise application of* T_t. *If, in addition, there exists a constant* $C_1 > 0$ *such that the Fourier transforms* $\hat{\chi}_l(\xi)$ *of the atoms* χ_l *satisfy the decay condition*

$$|\hat{\chi}_l(\xi)| \, |\xi| \leq C, \quad a.e. \ \xi \in \mathbb{R}^2, \text{ for all } l \in \mathbb{N}_0,$$

then

$$\left(\sum_{q \in \Lambda_1^l} \left\| \Phi_\Omega^l(T_t f) - \Phi_\Omega^l(f) \right\|_2^2 \right)^{1/2} \leq \frac{2\pi |t| C}{S_1 \cdots S_l} \|f\|_2,$$

for all $f \in L^2(\mathbb{R}^2)$ *and* $t \in \mathbb{R}^2$.

ii. *There exists a constant* $C_2 > 0$ *such that for all* $f \in L^2(\mathbb{R}^2)$ *that are band-limited,* $\omega \in C(\mathbb{R}^2, \mathbb{R})$, *and* $\tau \in C^1(\mathbb{R}^2, \mathbb{R}^2)$ *with Jacobian matrix* $D\tau$ *satisfying* $\|D\tau\|_\infty \leq 1/4$, *the feature extractor* Φ_Ω^l *for* $l \in \mathbb{N}$ *satisfies*

$$\left(\sum_{q \in \Lambda_1^l} \left\| \Phi_\Omega^l(F_{\tau,\omega} f) - \Phi_\Omega^l(f) \right\|_2^2 \right)^{1/2} \leq C_2 \left(R \|\tau\|_\infty + \|\omega\|_\infty \right) \|f\|_2.$$

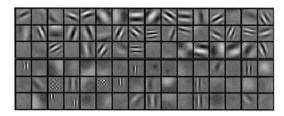

Figure 6.13: Visualization of the first layer of AlexNet [72].

6.2.2   Shearlet Initialized CNNs

In our work, we adapt some ideas of the approaches presented above. But instead of wavelets, we use shearlets since they are an extension of wavelets especially developed for the treatment of multivariate data such as images. Furthermore, Wiatowski and Bölcskei [123] provide a justification of our approach from a theoretical point of view. However, we do not initialize the first layers completely by shearlets but only a portion of it. We describe the reasoning for this procedure on the basis of Figure 6.13. It visualizes the filters which have been learned in the first convolutional layer of AlexNet [72]. In the original implementation, each layer of the network has been splitted in two parts and trained on two GPUs separately. Since the communication between the two GPUs only happens at particular locations in the network, the first half of the filter set is significantly different to the second half. For our work, we can find some valuable characteristics of the filters. It appears that roughly one half of the filters is used for edge detection. These filters are characterized by straight lines oriented in different directions. They show a close resemblance to shearlet filters, however they appear to be slightly noisy. Furthermore, one cannot recognize particular scalings. The rest of the filters in this layer mostly detect color combinations in the image, which cannot be realized by the application of shearlet filters.

For the generation of our *Shearlet Initialized CNN (SI-CNN)* based on AlexNet, we proceed as follows. First, we use the structure of AlexNet and initialize one half of the filters in the first layer by specific precomputed shearlet filters. To this end, we may use one or more shearlet systems. Then, we use a random distribution of the weights for the rest of the filters. Let Γ_1 denote the filter set of the first convolutional layer containing n_1 filters. If we use n_s shearlet systems $\Psi_1, \ldots, \Psi_{n_s}$ with a total number of η shearlets, we have

$$\Gamma_1 := \left\{ \Psi_1, \ldots, \Psi_{n_s}, \gamma_1^1, \ldots \gamma_{n_1-\eta}^1 \right\},$$

where $\gamma_1^1, \ldots \gamma_{n_1-\eta}^1$ are randomly initialized filters. Finally, we train this network from scratch. The usage of pretrained posterior layers does not appear useful since they depend on the filters of the first layer. In Section 6.3, we experiment with two options. We either fix the shearlet filters or we use them only for initialization but leaving them trainable. In the first case, one half of the first layer convolves the input image with shearlet filters. In fact, this portion of the first layer computes the shearlet transform of the input image. In 6.14, we illustrate our approach on the integration of shearlets in a CNN.

For our approach, the advantage of the AlexNet architecture is the filter size of 11×11 in the first convolutional layer. For this size, it is feasible to construct filters with local precision shearlets. Other available shearlet implementations such as the FFST [60, 61] and ShearLab

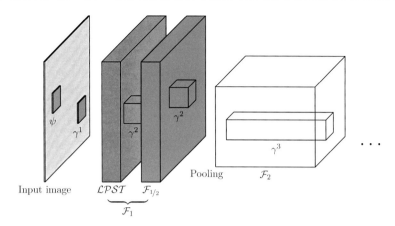

Figure 6.14: Illustration of the shearlet integration into a CNN. The filter ψ represents a shearlet whereas γ^1, γ^2 and γ^3 denote randomly initialized filters of the first, second and third convolutional layer. The feature map after the first convolutional layer $\mathcal{F}_1$ consists of the local precision shearlet transform and the convolution result with the filters γ^1, which is denoted by $\mathcal{F}_{1/2}$.

3D [77] create shearlet filters a lot bigger than 11×11. In contrast to AlexNet, the VGG16 network used for the SDS-RCNN algorithm [9] uses filters of size 3×3 in all convolutional layers. Even if we combine the first two convolutional layers before the first pooling layer, we still have a 5×5-sized receptive field with respect to the input image. This size is too small to create shearlet filters with adequate quality. Therefore, we have to adapt the network architecture. We use 9×9-sized filters in the first convolutional layer and directly attach the first pooling layer. Despite the increased filter size, we can still save training time if we fix the shearlet filters.

Since a training from scratch with a versatile data set like ImageNet consumes a lot of time, we train our networks with smaller data sets. As we lose accuracy thereby, we train the standard network corresponding to the respective SI-CNN with the same data sets. As a result, we obtain reference values which tell us if our approach is promising.

6.2.3   Implementation Details

For our implementation of CNNs, we use the open source framework *Caffe* [66]. This framework has been developed by the Berkeley Vision and Learning Center and is especially designed for computer vision tasks. A major advantage of Caffe is its modularization, which enables us to quickly adapt existing CNN models. In addition to it, the separation of the model definition with the protocol buffer language by Google and the actual implementation in C++ facilitates a rapid evaluation of new models. For more details about protocol buffers, see https://developers.google. com/protocol-buffers/. Furthermore, Caffe provides interfaces to Python and Matlab which can simplify implementations. In particular, we use the Matlab interface for the implementation of the SDS-RCNN algorithm. The Python interface is employed for the visualization of the filters in the first convolutional layer of our networks. The Python functions for this feature are available under https://github.com/smistad/visualize-caffe/.

Caffe stores all data (inputs, weights, gradients) in structures called *blobs*. These blobs are multidimensional arrays which store data for CPU as well as for GPU calculations. The transfer from CPU to GPU and vice versa is carried out by a `SyncedMem` class. Any value in a blob can be read as constant or as variable. Especially, the constant access is used to minimize the transfer between CPU and GPU.

The application of functions in Caffe is implemented by *layers*. A layer transfers one or more (bottom) blobs into one or more (top) blobs. Particular functions like the ReLU function can be calculated "in place", i.e. without an additional blob. Each layer implements a setup function as well as one backward pass and one forward pass for CPU and GPU respectively. The setup function initializes the network layers. The forward pass calculates the output of the network layers with input of the bottom blobs and delivers it to the top blobs. The network architecture can be specified simply by the definition of layers and corresponding bottom and top blobs in a `.prototxt` file.

Caffe uses so-called *weight filler* for the initialization of weights in a network layer. We make use of these weight fillers for the initialization of a CNN with shearlets. Since the shearlet transform is in fact a convolution, we use convolutional layers in Caffe and initialize them with precomputed shearlet filters. To this end, we have a closer look at the structure of a blob in Caffe, since the weights for the weight filler are also stored in blobs.

A blob for a convolutional layer has four dimensions (N, C, H, W), where N is the number of filters, C is the number of input channels, H is the height of the filters and W is the width of the filters in pixels respectively. Due to the internal representation of a blob as one-dimensional array, the value for index (n, c, h, w) is located internally at index $((nC + k) H + h) W + w$.

For the computation of the shearlet filters, we use our Matlab implementation of local precision shearlets already used in Chapter 5. With this implementation, we obtain a three-dimensional matrix $\mathcal{W}_{shear}$ with dimensions (H, W, N) containing one or more local precision shearlet systems. To consider the number of input channels, we introduce a fourth dimension C and use the same filters for each channel.

Now, the matrix $\mathcal{W}_{shear}$ has to be converted into a structure that a Caffe blob expects. For that, we make use of the indexing order described above. We have to perform an index shift since the indexing above is based on C++ arrays starting with index 0, whereas Matlab indexing starts with 1. Finally, we obtain a weight vector $\widetilde{\mathcal{W}}_{shear} \in \mathbb{R}^{NCHW}$ with

$$\widetilde{\mathcal{W}}_{shear}\left(\left(\left(\left(n - 1\right)C + c - 1\right)H + h - 1\right)W + w\right) = \mathcal{W}_{shear}\left(h, w, n, c\right),$$

with $h = 1, \ldots, H$, $w = 1, \ldots, W$, $n = 1, \ldots, N$ and $c = 1, \ldots, C$. Now, we export this vector into a text file in which row i contains the weight that shall initialize the blob at position $i - 1$. Finally, we can easily read this text file via C++ and integrate the corresponding function in the Caffe header file `filler.hpp`.

6.3   Experiments

We split our experiments in two parts. First, we deal with the more simple task of pedestrian classification. We use a modest training data set to speed up the training process and thus to quickly evaluate different settings. Second, we apply our insights of the first step to the more complex task of pedestrian detection. For that purpose, we use the SDS-RCNN algorithm [9] presented in Section 6.1.3 and exchange the pretrained filters with our own ones.

For the initialization of the first layer with shearlets, we use local precision shearlet systems. Deduced from our experiments for the pedestrian detection with hand-crafted features in Section 5.5, we always use three scales, i.e. $j_0 = 3$. As indicated before, further scales reduce the shearlet support to such an extent that they are useless for practical application. Furthermore, we use the best performing degree of anisotropy in Section 5.5, i.e. $\alpha = 3/4$. We mainly use 8 shearlets per scale, since this number is most suitable to attain the needed number of filters for the shearlet initialization. Only when we obtain too many filters, we partly use 6 shearlets per scale. The scope for development is mainly concerned with the spline order p and the number of derivatives q of spline shearlets. We consider spline shearlets with $q = 1$ since they show the best results for hand-crafted feature detectors, but also higher derivatives. These shearlets show a strong similarity to the AlexNet filters in Figure 6.13. We remark that our filters do not represent the complete shearlets at coarse scales. Small values at the border are "clipped off". In this way, our filters correspond to the learned filters in Figure 6.13. To change the size of shearlet within a filter, we adapt the values of p and q as well as the sampling constant c.

6.3.1 Pedestrian Classification

For our first experiments, we use the AlexNet architecture. We initialize 48 filters of the first layer with shearlets. For the remaining 48 filters, we use a random, Gaussian initialization with mean 0 and standard deviation 0.01. Furthermore, we initialize the bias with 0. Except the shearlet initialization, this corresponds to the standard initialization of AlexNet. On one side, we evaluate the results if we fix the shearlet filters such that only the random initialized filters are learned. In comparison, we evaluate what happens if we enable the shearlet filters for the learning procedure.

Concerning the data set, we use samples of the Caltech data set with size 128×64. For training, we use around $2 \cdot 10^5$ images for class 0, i.e. "no pedestrian", and around $5 \cdot 10^4$ images for class 1. All images are sampled from the training portion of the Caltech data set. For test, we sample $4,000$ images for class 0 and $1,000$ for class 1 from the Caltech test data set. This amount of images is sufficient for basic tests on the effect of the shearlet initialization.

Turning to the shearlet parametrization, we use five variants $i \in \{1, \ldots, 5\}$. For the first four variants, we apply $(p_1, q_1) = (3, 1)$ and $(p_2, q_2) = (7, 3)$. With 8 shearlets per scale, we get $2 \cdot 24 = 48$ shearlet filters as required. The three variants distinguish themselves by the choice of different sampling constants. We denote the vector of sampling constants for variant i by $c^i = (c_1^i, c_2^i)$, where c_j^i is the sampling constant for (p_j, q_j). We set

$$
\begin{aligned}
c^1 &= (0.25, 0.35), \\
c^2 &= (0.30, 0.40), \\
c^3 &= (0.35, 0.45), \\
c^4 &= (0.45, 0.75),
\end{aligned}
$$

such that the shearlets are represented smaller with each variant. Only with variant 4, the shearlets of the coarsest scales are represented completely in the corresponding filters. This filter variant is visualized in Figure 6.15. Each filter is shown three times, since we apply each filter to all color channels of RGB images.

Finally, we consider a fifth variant which uses only shearlet filters and no randomly initialized ones. We use filters of variant 2 and in addition shearlets with $(p_3, q_3) = (7, 5)$ and $(p_4, q_4) = (5, 3)$, whereas we only use 6 shearlets per scale. Furthermore, we compute a low-pass filter for each (p, q)-pair to have a closer correspondence to the original AlexNet filters. To reach the

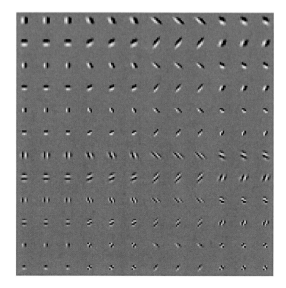

Figure 6.15: Visualization of the shearlet filters according to variant 4.

required number of 96 filters, we include shearlets with just one scale and 8 shears. To be very different from the remaining filters, we use $(p_5, q_5) = (11, 9)$.

Now, we train our five variants with fixed shearlet filters on the data set described above. We use a maximum number of $40,000$ iterations with a batch size of 128 and a learning rate of 0.01. Table 6.1 shows the best classification result for each variant on the test data set itemized by classes as well as the result after only $5,000$ iterations. The percentages state the proportion of correctly classified images of each class. We remark that variant 4 has been realized at a later stage of our work. Therefore, results for this variant are not shown in the table. We recognize a clear advantage of SI-CNNs at an early stage of training after $5,000$ iterations. This speaks for the utilization of an initialization with shearlets. Throughout the training, the randomly initialized networks catch up but do not reach the SI-CNN results completely. Furthermore we can see that SI-CNNs reach their best results earlier than randomly initialized networks.

The results of all networks show a better classification of class 0. Therefore, we adapt the training data set. To increase the proportion of pedestrians, we remove one half of the images for class 0 for the training set as well as for the testing set. Subsequently, we flip each image to obtain a bigger amount of images. In this way, we obtain a training data set of around $2 \cdot 10^5$ images of class 0 and $1 \cdot 10^5$ images of class 1. The test data set then contains around $2,000$ images of class 0 and $1,000$ images of class 1. Table 6.2 shows the results governed with the new data set, whereas we illustrate the results at an early stage of the training already after $1,000$ iterations. Results of variant 4 after $1,000$ iterations are missing, again because it has been developed at a later stage of our work.

We can recognize differences at an early stage of the training very clearly now. After $1,000$ iterations, a randomly initialized network classifies every test image as "no pedestrian". In comparison, SI-CNNs deliver useful results already at this stage. Again, the results get more

Variant	Iteration	Class 0	Class 1	Overall
Gaussian	5,000	98.86%	89.16%	96.91%
Gaussian	40,000	99.72%	95.08%	98.79%
1	5,000	98.91%	96.69%	98.47%
1	30,000	99.62%	96.39%	98.97%
2	5,000	99.37%	96.39%	98.77%
2	20,000	99.49%	96.08%	98.81%
3	5,000	98.23%	99.10%	98.41%
3	10,000	99.57%	96.49%	98.95%
5	5,000	99.47%	96.08%	98.79%
5	20,000	99.49%	96.49%	98.89%

Table 6.1: Classification results after the first training.

Variant	Iteration	Class 0	Class 1	Overall
Gaussian	1,000	100%	0%	67.02%
Gaussian	25,000	99.25%	97.36%	98.63%
1	1,000	99.35%	92.70%	97.16%
1	15,000	99.10%	97.46%	98.56%
2	1,000	98.90%	91.08%	96.32%
2	20,000	99.10%	97.77%	98.66%
3	1,000	99.20%	91.99%	96.82%
3	20,000	99.15%	98.17%	98.83%
4	10,000	98.95%	99.09%	99.00%
5	1,000	98.45%	96.15%	97.69%
5	25,000	99.10%	98.07%	98.76%

Table 6.2: Classification results after the second training.

similar throughout the training. Still, the best performing network is a SI-CNN, more precisely variant 4. Furthermore, we recognize that randomly initialized networks show a more unequal treatment of the two classes. Images of class 1 are misclassified significantly more often than images of class 0. Finally, we see that variants with a bigger value of the sampling constant perform better.

Now, two questions arise. First, since we have fixed the shearlet filters until now, which results will we achieve if we enable them for training? Second, what will happen if we train with a maximum number of iterations bigger than 40,000 iterations?

First, we consider variant 5 which is the only one that uses shearlet filters only. Again, we train with a maximum number of 40,000 iterations but with learnable shearlets now. The best result is achieved after 20,000 iterations with a classification rate of 99.15% for class 0, 96.96% for class 1 and 98.43% overall. This result is inferior to the one with fixed filters. Therefore, we will not evaluate variant 5 during further experiments. Furthermore, we will not use variants 1 and 2 since Table 6.2 reports inferior results for them. Now, we train the remaining variants 3 and 4 with a maximum of 350,000 iterations and use fixed and learnable shearlet filters respectively. We multiply the learning rate of 0.01 by 0.1 each 100,000 iterations, as recommended in the Caffe tutorial[1]. In addition, we train our reference network with Gaussian initialization with the same settings. Table 6.3 reports the classification results after this training.

---

[1]http://caffe.berkeleyvision.org/tutorial/

Variant	Iteration	Class 0	Class 1	Overall
Gaussian	$350,000$	99.45%	96.55%	98.49%
3 / fixed	$350,000$	99.30%	97.67%	98.76%
3 / learnable	$350,000$	99.50%	97.06%	98.70%
4 / fixed	$350,000$	99.40%	97.67%	98.83%
4 / learnable	$350,000$	99.55%	96.96%	98.70%

Table 6.3: Classification results after the third training.

For both variants of shearlet initialization, we achieve better results in case they are fixed and not learnable. Again, variant 4 shows up to be more successful than variant 3. Both shearlet networks provide significantly better results than the randomly initialized network, especially for class 1.

As a final test concerning pedestrian classification, we use an externally pre-trained AlexNet and compare its performance to the one of our SI-CNNs. Such models can be downloaded from the so-called Model Zoo of Caffe under https://github.com/BVLC/caffe/wiki/Model-Zoo. We train this model on our data set for $150,000$ iterations with a learning rate of 0.01 and then multiply the learning rate by 0.1 for a training of further $150,000$ iterations. The best result is achieved after $20,000$ iterations with classification rates of 99.35% for class 0, 96.55% for class 1 and a total of 98.83%. Therefore, without any pre-training on diverse training data sets, we achieve the same result with the SI-CNN of variant 4 as an extensively pre-trained network.

6.3.2   Pedestrian Detection

During our experiments concerning pedestrian classification, we have noticed that the shearlet initialization of parts of the first layer has a positive effect on the classification results. Therefore, we want to transfer our approach to the task of pedestrian detection. As indicated before, we use the state-of-the-art algorithm SDS-RCNN [9]. In this approach, features are extracted by a pre-trained VGG16 network. But until now, we do not have shearlet initialized VGG16 networks available. Therefore, we first use AlexNet as a basic structure for feature extraction.

We utilize the AlexNet SI-CNN of variant 4 which we have trained with a maximum of $350,000$ iterations. According to the insights of the previous section, we always fix the shearlet filters. As a reference network, we use the Gaussian initialized network trained with $350,000$ iterations as well as the AlexNet model from the Caffe Model Zoo.

Originally, the filters of the convolutional layers before the second pooling layer are fixed during the training of the SDS-RCNN algorithm. Since our pre-training is not as extensive as in the original approach, we set these first filters, except the shearlet filters, as learnable. We take over all other settings from the original SDS-RCNN algorithm. Especially, we start with a learning rate of 0.001 and multiply it by 0.1 after $60,000$ iterations and stop the training after $120,000$ iterations.

The SDS-RCNN implementation delivers three log-average miss rates. The result when RPN is used exclusively, the result with exclusive utilization of R-CNN and finally the result when the classification probabilities are combined according to the procedure described in Section 6.1.3.

Table 6.4 shows the results of the SDS-RCNN application with different CNNs based on AlexNet. The SI-CNN-v4 provides better results than the randomly initialized CNN. But the results are significantly inferior to the rates of the AlexNet from the Model Zoo. We deduce that our pre-training is not sufficient. However, the AlexNet from the Model Zoo also provides results in

Variant	RPN	R-CNN	Fusion
Gaussian	62.83%	40.85%	38.89%
SI-CNN-v4	53.44%	39.54%	36.10%
Model Zoo	33.47%	26.11%	22.97%

Table 6.4: SDS-RCNN results with AlexNet models.

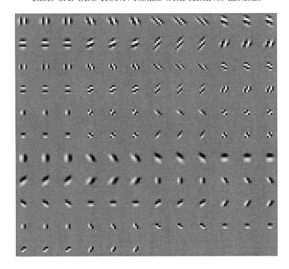

Figure 6.16: Visualization of 9×9 sized shearlet filters for VGG16.

great need of improvement. The combined log-average miss rate of 22.97% is significantly higher than the one we achieved with hand-crafted shearlet features in Section 5.5. We conclude that AlexNet is not sufficient as the basis for the feature extraction.

Therefore, we use only the VGG16 architecture during our remaining experiments. As described in Section 6.2, the filter size of the original VGG16 is too small for a straightforward shearlet initialization. Our resolution is to utilize one layer with 9×9 sized filters instead of the first two layers with filters of size 3×3. We split the 64 filters of this layer into 42 shearlet filters and 22 randomly initialized filters. Concerning the shearlet filters, we use 8 shears per scale with $(p_1, q_1, c_1) = (5, 3, 0.65)$ as well as 6 shears per scale with $(p_2, q_2, c_2) = (3, 1, 0.45)$. The corresponding filters are visualized in Figure 6.16.

Furthermore, we use a second network with filters of size 7×7 instead of 9×9 in the first layer. For these filters, we use the same spline parameters but use 6 shears for the first spline and 8 for the second one. For the sampling constants we use $c_1 = 0.85$ and $c_2 = 0.6$. The choice of these sampling constants follows the observation from the previous section that we achieve better results if the shearlets are not clipped off.

During our previous pedestrian detection experiments, we found that our pre-training was not sufficient. Especially, our training data set only contained two classes so far. Therefore, we want to use a more comprehensive data set to be able to extract more general features. We choose the CIFAR-100 data set, which has been introduced in [71] and is online available under

Variant	RPN	R-CNN	Fusion
Reference / 3×3	17.91%	24.65%	14.73%
Reference / 9×9	18.51%	24.22%	14.76%
SI-CNN / 9×9	15.93%	24.59%	13.74%
SI-CNN / 7×7	21.42%	25.79%	17.01%

Table 6.5: SDS-RCNN results with VGG16 and pre-training with CIFAR-100.

https://www.cs.toronto.edu/~kriz/cifar.html. The CIFAR-100 data set consists of 100 classes with 600 images respectively. The images have size 32×32, whereby the fully-connected layer of our network is built by relatively few neurons. This speeds up our training. Furthermore, the data set contains images of persons, which may be beneficial for our task of pedestrian detection.

Unfortunately, the training of VGG16 with the CIFAR-100 data set yields some difficulties. First, VGG16 is a relatively deep network, such that no convergence can be achieved with Gaussian initialization. Simonyan and Zisserman [113] solved this problem by training more shallow networks with random initialization and then using the trained filters for initialization of the target network. We choose a different procedure and directly train the deep network with the initialization method described by He et al. [62]. Instead of setting the standard deviation of the Gaussian distribution fixed to 0.01, we count the number of inputs of a neuron n and then set the standard deviation to $\sqrt{2/n}$. This type of initialization results in a good convergence behavior, especially for neurons with the ReLU activation function.

Another issue we face is overfitting caused by the relatively small training data set of $50,000$ images. As a resolution, we first augment the data set by flipping each image. Second, we perform a normalization over the training batches after each layer. This method is called a *batch normalization* and introduced by Ioffe and Szegedy [65].

Finally, we remember that the first two fully-connected layers of VGG16 are used during the training of the classification network of the SDS-RCNN algorithm. We cannot take over this procedure, since we would need a training on images of size 224×224. Therefore, we initialize our fully-connected layers randomly for our training of SDS-RCNN.

Since we changed the architecture of VGG16, we use two reference networks. The first reference network uses the original VGG16 with 3×3 filters in the first layer, the second one uses 9×9 sized filters. We train all networks with a maximum of $350,000$ iterations with a learning rate of 0.01 that we multiply by 0.1 each $100,000$ iterations. The results are reported in Table 6.5.

We can see that the miss rates are significantly better than the ones with AlexNet. Since the SI-CNN with 7×7 sized filters performs significantly worse than all other networks, we will not consider this variant during our remaining experiments. Concerning the reference networks, we do not find a significant difference in the detection quality if we use the original VGG16 architecture or the adapted one. However, the original architecture is more appealing since it requires less memory. The SI-CNN with 9×9 sized filters performs about 7% better in relation to the reference networks. However, this improvement appears mainly in the RPN part.

In summary, we were able at all times to improve the classification and detection rates by a shearlet initialization compared to corresponding reference networks. However, the achieved log-average miss rate of 13.74% of a SI-CNN based on VGG16 and pre-trained on CIFAR-100 is clearly inferior to the one reported by Brazil et al. [9]. By the utilization of the original VGG16 pre-trained on the enormous ImageNet data set [24] a log-average miss rate of 7.36% is reached. We finally conclude that shearlets can improve neural networks but they cannot ease the need of huge, comprehensive data sets for the training of CNNs. We leave the extensive trainings of SI-CNNs with the ImageNet data set as a perspective for further research.

6.4 Conclusion

In this chapter, we evaluated in which way the initialization of the first layer of a CNN with shearlets affects the performance of the network for the classification and the detection of pedestrians. Our basic idea is that early CNN layers intuitively perform an edge detection. From a theoretical point of view, shearlets provide optimal filters for this task. To merge the theory with procedures which have shown to deliver best results in practice, we initialize a portion of the first layer filters by shearlets. The remaining filters are initialized and learned as in the corresponding state-of-the-art procedures.

In our first experiments, we addressed only the classification of pedestrians. Here, we found that CNNs initialized with shearlets in the first layer show satisfactory results already after very early stages of the training. In comparison, randomly initialized networks can not provide useful results after the same training stage. Furthermore, we measured that classification results are always better if we fix the shearlet filters and do not enable them for training. During all classification experiments, we achieved better results with shearlet initialization than with corresponding reference networks.

In further experiments, we use these insights for the task of pedestrian detection. First, we noticed that the shearlet initialization cannot improve relatively shallow networks as AlexNet [72] such that they could be applied for the detection of pedestrians with state-of-the-art results. Concerning deeper architectures, we were able to improve the detection performance of the VGG16 model [113] by shearlet initialization trained on the CIFAR-100 data set. The original VGG16 trained on the same data and with the same parameters cannot reach the results using shearlets. However, the best result of the SI-CNN trained on CIFAR-100 is clearly inferior to the one achieved with the utilization of the VGG16 pre-trained on the extensive ImageNet data set as reported in [9]. This finding underpins the immense power of data for object detection algorithms.

We conclude that shearlet initialization cannot ease the need of a pre-training with huge and comprehensive data sets. Such a training requires a vast amount of time and specific hardware. Considering one wants to estimate the behavior of a CNN with a new network architecture a large-scaled training is not mandatory. Parameters can be adjusted more quickly if the training is firstly restrained to smaller data sets. In this case the shearlet initialization can be very beneficial, since we measured a better convergence behavior at the beginning of the training. As a perspective for further research, it has to be evaluated in detail if the shearlet initialization still leads to improvement in case of a training with an extensive data set such as ImageNet.

"What really matters is what you do with what you have."

H.G. Wells

7

Embedded Realization

A major application area of pedestrian detection is the automotive sector. Currently deployed warning or brake intervention systems in vehicles profit from an accurate pedestrian detection to potentially save numerous lives. However, to be applied in a vehicle, a detection algorithm has to be runable with restricted hardware requirements of an automotive Electronic Control Unit (ECU), i.e. an embedded system. In addition, it has to provide a useful frame rate on this hardware system.

In this chapter, we analyze in which way the local precision shearlet transform and a pedestrian detection algorithm based on it can be realized on an embedded system. Therefore, we re-implement our base detection algorithm shearFtrs-v1 in C++ programming language whereas the NVIDIA Jetson TK1 developer kit [98] serves as an embedded target. As architectural foundation the *Robot Operating System (ROS)* [100] is used, which is placed on a Ubuntu operating system.

7.1 Hardware System

The developer kit Jetson TK1 from NVIDIA provides a platform, which uses the performance of a graphic processor for the area of embedded systems and its tasks. It utilizes a NVIDIA Kepler GPU with 192 CUDA Cores in combination with Quad-Core ARM Cortex A15 CPU and 2 GB RAM. The NVIDIA Jetson contains a wide range of conventional periphery which makes it useable for many applications. The kit includes

◇ 16 GB eMMC Memory

◇ 1 Full-Size HDMI Port

◇ 1 RTL8111GS Realtek GigE LAN

◇ 1 SATA Data Port

The complete content of the developer kit can be taken from its user guide [97]. In addition to the platform, NVIDIA delivers a wide range of tools such as the IDE NSight or the analysis tool Tegra System Profiler.

Figure 7.1: NVIDIA Jetson TK1 developer kit [98].

7.2   Software System

A main focus for the choice of the software system is that it shall have high modularity and easy extensibility. During research for suitable and available architectures, the *Robot Operating System (ROS)* [100] and the *Microsoft Robotics Developer Studio (MRDS)* [94] appear. Our hardware system and its development tools require a Ubuntu operating system. Since it is set up on Ubuntu, we choose ROS as a basis for our software system. In 2007, Willow Garage and the University of Stanford began the development of the Robot Operating Systems. It is a mixture of OS and Middleware, whereas it is similar to a service oriented architecture. ROS is based on a peer-to-peer communication connected to a buffer and a lookup system, which enables to communicate with every process synchronous or asynchronous. With a microkernel design based on multiple tools for build and run processes of the different components, ROS achieves a high degree of isolation of each command. That means each command is an isolated execute such that failures only influence the corresponding process and not the complete system. ROS supports different types of programming languages, for example Python or C++. The peer-to-peer communication is processed via XML-RPC, which is available in most of the programming languages. The libraries of ROS are set up on the Ubuntu OS and are located above the layer of device drivers, Ubuntu scheduling and file system. Thus, the Ubuntu scheduler directly influences the ROS processes.

The basic principle of ROS is the parallel processing of a high amount of executes and the synchronous or asynchronous exchange of information. The executes are mapped to *nodes*, whereas each node subscribes itself via TCP/IP or UDP to the *ROS master* and delivers its configuration. The master can be seen as a broker of the complete system. It is a declaration and memory service, enabling the nodes to know each other and to communicate. Furthermore, the master provides a parameter service. With this service, parameters and data can be stored and retrieved flexibly. Thus, the ROS master fulfills the secondary function as a parameter database. The communication between the nodes is performed via *topics* in the asynchronous case and via *services* in the synchronous case. The communication via topics can be seen as an asynchronous information bus. One or more nodes can write information on a topic, whereas one or more nodes can retrieve this information. Services provide a synchronous request-answer interaction. In each case, a communication medium is needed. Therefore, ROS provides *messages* composed of a combination of kernel elements, such as integers or floats, and a *container*, which contains the kernel elements.

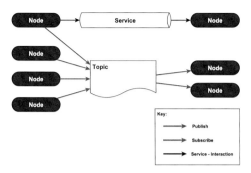

Figure 7.2: ROS communication.

7.3 System Architecture

The system architecture of our base detection algorithm is divided into ROS nodes. Each of these nodes executes specific tasks and provides services according to them. Besides services, the system also includes a publish/subscribe mechanism. This mechanism allows a node to subscribe a specific topic in which another node is publishing. The partitioning in nodes modularizes the system on a high level and allows easy extensions. Nodes are independent by each other and can be executed on different hardware platforms at the same time. For example, one can execute all computations on the embedded platform while the visualization is done on an external device. A specialty of this system is the concept of action-servers, which can be seen as an abstraction of services. In some cases, services consume relatively high runtime, e.g. for huge computations. In these situations, the system does not know in which state the service is. Furthermore, the service can not be terminated simply. For such cases, we design action-servers, which are part of the ROS package `actionlib`.

Our system contains 7 nodes, which are independent executables. That means, the execution of nodes is not affected from each other. In case of a failure situation on a node, the system is not affected in total but only the respective process. Our nodes communicate via messages, which are mostly included in service or action requests. Figure 7.3 shows an overview of our system where nodes are illustrated as gray components. These nodes are connected via interfaces by which they communicate with each other. Some nodes have special interfaces, which are illustrated in red in Figure 7.3. These interfaces carry out special communication, e.g. to an external device.

The system is composed of following nodes.

⬦ PD_Initialization_Node:
  The PD_Initialization_Node computes mainly the shearlet filters according to the description in Chapter 3. They are provided to the remaining nodes in a topic via the publisher pub_PD_Initialization_PShearlets.

⬦ PD_Detection_Node:
  Given an input image provided by the PD_InputOutput_Node, the PD_Detection_Node carries out the pedestrian detection.

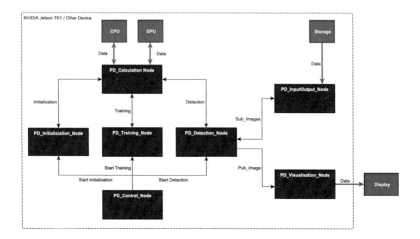

Figure 7.3: System architecture of our detection algorithm.

◇ PD_InputOutput_Node:
This node is the interface to the system environment and the stored data sets. The latter can be loaded and be provided via a topic.

◇ PD_Control_Node:
The PD_Control_Node takes over the control of the complete system.

◇ PD_Training_Node:
This node contains the training of the AdaBoost classifiers. All relevant training data is loaded and subsequent calculations are initiated.

◇ PD_Calculation_Node:
The PD_Calculation_Node carries out all fundamental calculations. These calculations are provided via different action-servers.

◇ PD_Visualization_Node:
This node displays the detection results provided by the system.

7.3.1   Interfaces

To ensure the communication between the abovementioned nodes, we have to set up a precise interface description. We will use ROS messages, ROS services and ROS actions as interfaces. These communication concepts are provided by the ROS architecture. Figure 7.4 shows an overview over the communication between the nodes. In the following, we will provide a complete description of the interfaces in our system.

Messages

The concept of ROS messages is the fundamental building block of the communication in our system. The system contains 47 messages and uses these messages in all interactions between the

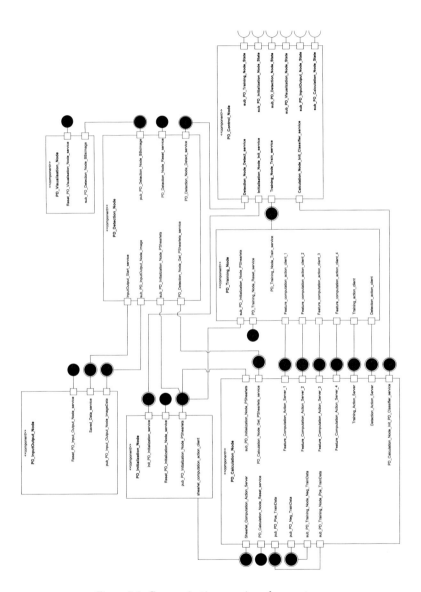

Figure 7.4: Communication overview of our system.

Standard type	Unit	C++
bool	unsigned 8-bit int	uint8_t
int8	signed 8-bit int	int8_t
uint8	unsigned 8-bit int	uint8_t
int16	signed 16-bit int	int16_t
uint16	unsigned 16-bit int	uint16_t
int32	signed 32-bit int	int32_t
uint32	unsigned 32-bit int	uint32_t
int64	signed 64-bit int	int64_t
uint64	unsigned 64-bit int	uint64_t
float32	32-bit IEEE float	float
float64	64-bit IEEE float	double
string	ascii string (4)	std::string
time	secs/nsecs signed 32-bit ints	ros::Time
duration	secs/nsecs signed 32-bit ints	ros::Duration

Table 7.1: Standard types of ROS Messages

nodes. Messages are available as ROS standard types and stored in .msg files. The available types are shown in Table 7.1. In addition, arrays of these standard types are allowed. A special case is the message type `sensr_msgs/Image`, which is contained in the ROS package `Sensor_Msg`. This message type is used for the transfer of images in a ROS system. In this chapter, the computer vision library OpenCV is used. Therefore, the content of `sensr_msgs/Image` message has to be converted to a OpenCV compatible format. This task is carried out via the ROS package `CV_Bridge`. For a detailed description of the abovementioned packages, see [100].

For any of our 47 messages, classes and functions are available to convert a ROS datatype into an OpenCV datatype and vice versa. Exemplary, Listing 7.1 shows function `toMsg` for the message `pShearlets`. This function converts the data content of the message class and stores it in the ROS message `pShearlets`. This specific message is implemented by the object `embedded_pd::PShearlet pShearlets`. In line 22, we have the conversion of an OpenCV image to a `sensr_msgs/Image`. The function `MatToMsg` will be described more detailed in Section 7.4. The message `pShearlets` is part of an action-server or a service request and serves as a container for the calculated shearlet parameters.

Topics

Topics are used to identify messages and to control the communication. A node writes messages to topics such that another node can read this message via the corresponding topic. The communication via topics can be seen as an asynchronous interaction paradigm. A node that has subscribed a topic receives a notification if new data is available for this topic.

In our pedestrian detection system, we have 5 topics with 5 publishers which are writing messages and 7 subscribers which are reading them. Exemplary, the publisher `pub_PD_Initialization_PShearlets` writes the data of the shearlet parameters on the topic `PD_Initialization_Node/PShearlets`. Correspondingly, the subscriber `sub_PD_Initialization_Node_PShearlets` reads this data. The topic structure in this example is illustrated in Figure 7.5.

```
embedded_pd::PShearlets PD_Util::PShearlets::toMsg()
{
    sensor_msgs::Image result_image;
    cv_bridge::CvImage cv_image;
    embedded_pd::PShearlets pShearlets;

    pShearlets.nameOfShearlet = this->nameOfShearlet;
    pShearlets.dim_img = this->dim_img;
    pShearlets.supp = this->supp;
    pShearlets.shearsPerScale = this->shearsPerScale;
    pShearlets.numOfAllShears = this->numOfAllShears;
    pShearlets.nScales = this->nScales;
    pShearlets.scalesUsed = this->scalesUsed;
    pShearlets.RMS = this->RMS;
    pShearlets.ShearletIdx = this->shearletIdxs;
    pShearlets.thresholdingFactor = this->thresholdingFactor;
    pShearlets.flag_fourier = this->flag_fourier;

    ros::Time time = ros::Time::now();
    for (std::size_t i = 0; i < Psi.size(); i++)
    {
        result_image = PD_Util::MatToMsg(this -> Psi.at(i),"Psi", i, sensor_msgs::
            image_encodings::TYPE_32FC1);
        pShearlets.Psi.push_back(result_image);
    }
    return pShearlets;
}
```

Listing 7.1: Conversion struct to PShearlets Message

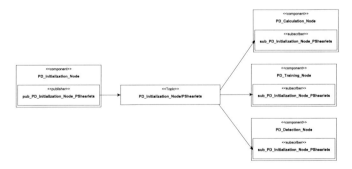

Figure 7.5: Topic structure of message PShearlets

```
1  #request
2  string text_in
3
4  ---
5
6  #response
7  PShearlets pshearlets
```

Listing 7.2: Service definition for Get_PShearlets.

Services

Our system contains 12 services, which are provided by the nodes. Services are defined in .srv files and compiled to source code by the *ROS Client Library*. The format of service definitions is based on the one for messages and is extended by a *request/response* functionality. As can be seen in Listing 7.2, the sections for request and response are separated by three consecutive '-'. Here, the definition of the service Get_PShearlets is shown as an example. This service is provided by the PD_Calculation_Node and is used to communicate the current shearlet parameters.

Actions

Our system contains 4 actions, which are exclusively provided by the PD_Calculation_Node. The reason behind it is that the PD_Calculation_Node executes all fundamental calculations and that actions are designed for long term calculation periods. In contrast to services, actions can be used synchronously and asynchronously. Actions are defined in .action files, while their format is similar to the one of services. Listing 7.3 exemplarily shows the definition of the action Shearlet_Computation. It is separated into 3 sections, i.e. goal, result and feedback. Each section contains message standard types as well as more complex messages, which are composed of different standard types. The content of the sections can be described as follows:

◇ goal
A goal is sent from the action-client to the action-server. It contains data to achieve the goal for which the action-server has been defined.

◇ feedback
A feedback is sent from the action-server to the action-client and delivers the progress of the calculation. This information can be sent several times.

◇ result
A result is sent from the action-server to the action-client when the corresponding goal is achieved. In contrast to a feedback, a result can only be sent once.

The action Shearlet_Computation is executed by the shearlet_computation_action_client of the PD_Initialization_Node, such that the shearlets are computed according to the data contained in the corresponding goal.

Parameter Server

The parameter server provides a platform where all parameters can be saved. It is offered by the ROS master, which serves as a central interface in our system. Parameters can be up- or downloaded via topics.

```
 1  # goal definition
 2  int32 shearlet_type
 3  int32 shearlet
 4  int32 [] dim_image
 5  int32 [] shearsPerScale
 6  int32 numOfAllShears
 7  int32 nScales
 8  float32 [] supp
 9
10  ---
11
12  # result definition
13  PShearlets pShearlets
14
15  ---
16
17  # feedback
18  int32[] sequence
```

Listing 7.3: Action definition for `Shearlet_Computation`.

7.3.2 Dynamic Behavior

In this section, we describe the dynamic behavior of our system with sequence diagrams and state machines. Sequence diagrams illustrate the communication in a system time-wise. Elements in such a diagram include communication partners, interactions, lifeline and messages [64].

Sequences

In the following, we will show two sequence diagrams which illustrate the behavior of our system. We will use the UML notation given in Table 7.2.

Sequence 1 - Procedure without Classifier File   The first sequence, shown in Figure 7.6, illustrates the procedure of the system with training of a classifier. The PD_Control_Node sends a message to the PD_Initialization_Node to initialize the shearlet parameters. Next, the PD_Initialization_Node sends a service request to the PD_Calculation_Node, which computes the corresponding shearlets. These are then sent back to the PD_Initialization_Node, which subsequently writes the shearlet parameter on a topic via an asynchronous message. Out of this topic, the PD_Training_Node and the PD_Detection_Node receive a message containing the shearlet parameters. Subsequently, the PD_Control_Node receives a message that the initialization is complete. Following, the training process is initiated by a message sent from the PD_Control_Node to the PD_Training_Node. This is the main difference between the two sequence diagrams. The PD_Training_Node sends a message to the PD_Calculation_Node to calculate the features for the training data. When the calculation is complete, the PD_Training_Node receives a positive response message. Subsequently, this node sends a message again to the PD_Calculation_Node, now to train the classifier. Once more, the PD_Training_Node receives a positive response when the calculation is finished. At this stage, all inputs for performing the detection task are available. To start the detection, the PD_Control_Node sends a message to the PD_Detection_Node. In turn, this node sends a message to the PD_Input_Output_Node, which loads one or more input images and writes it on a topic. The PD_Input_Output_Node sends a positive message that the loading process has been started. Next,

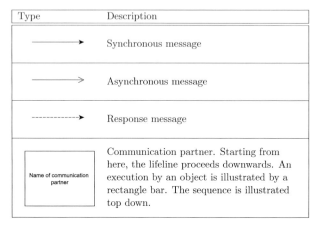

Type	Description
——————▶	Synchronous message
——————⟶	Asynchronous message
·············▶	Response message
Name of communication partner	Communication partner. Starting from here, the lifeline proceeds downwards. An execution by an object is illustrated by a rectangle bar. The sequence is illustrated top down.

Table 7.2: Sequence diagram notation.

the PD_Detection_Node receives a message with the input images and computes the image features via the PD_Calculation_Node. Once the features are computed, the classification is started by a message of the PD_Detection_Node to the PD_Calculation_Node. In response to that, the PD_Calculation_Node sends the detection results back to the PD_Detection_Node, which communicates them on a topic. This topic is read by the PD_Visualization_Node to display the detection results.

Sequence 2 - Procedure with Classifier File The second sequence, shown in Figure 7.7, illustrates the procedure in case no training prior to detection is needed. The only difference to the first sequence is that a pretrained classifier file is available. This file is loaded after request by the PD_Control_Node to the PD_Calculation_Node. Subsequently, detection and all further processes described in the first sequence are performed.

System States

In addition to sequence diagrams, we describe the behavior of the system with state machines. They enable us to illustrate the different states of an object in the system and its possible transitions. The state machines in this section are designed according to the modeling concept of David Harel [59]. As described earlier, our detection system consists of 7 nodes. Each node is started by the execution of a launch file, which is described more detailed in Section 7.4. The state machine of the system startup is illustrated in Figure 7.8. Here, the state transition of each node in its init state takes place.

Each node contains an individual state table with a mapping of states to corresponding functions. An example is given in Listing 7.4, which shows the state table for the PD_Initialization_Node. The state table of a node is initialized by a transition table, see for example Listing 7.5.

The state transition is performed by a simple function call, which is illustrated in Listing 7.6. In this function, the variable newState gets the corresponding enumeration value which causes the transition of the state machine. In the following, we describe the behavior of each node by its state machine.

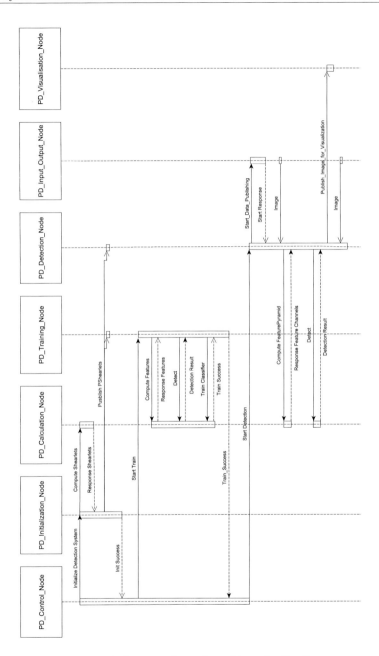

Figure 7.6: Sequence of procedure without classifier file.

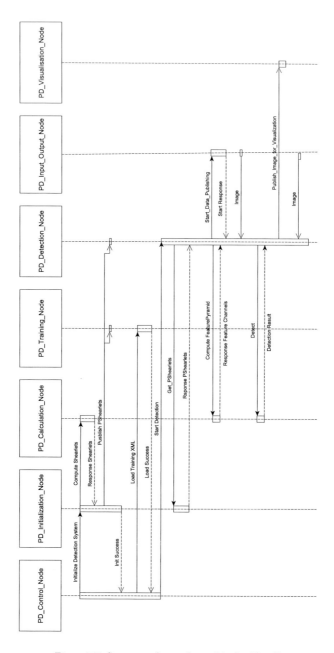

Figure 7.7: Sequence of procedure with classifier file.

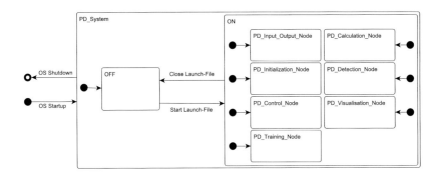

Figure 7.8: State machine for the system startup

```
1  BEGIN_STATE_MAP
2     STATE_MAP_ENTRY(&PD_Initialization_Node::PD_Init_Off)
3     STATE_MAP_ENTRY(&PD_Initialization_Node::PD_Init_On)
4     STATE_MAP_ENTRY(&PD_Initialization_Node::PD_Init_Publish_PShearlets)
5  END_STATE_MAP
```

Listing 7.4: State table for PD_Initialization_Node.

```
1  enum E_States
2  {
3     PD_INITIALIZATION_NODE_OFF = 0,
4     PD_INITIALIZATION_NODE_ON ,
5     PD_INITIALIZATION_NODE_PUBLISH_PSHEARLETS ,
6     ST_MAX_STATES
7  };
8
9  BEGIN_TRANSITION_MAP
10    TRANSITION_MAP_ENTRY(PD_INITIALIZATION_NODE_OFF)
11    TRANSITION_MAP_ENTRY(CANNOT_HAPPEN)
12    TRANSITION_MAP_ENTRY(CANNOT_HAPPEN)
13 END_TRANSITION_MAP(NULL)
```

Listing 7.5: Transition table for PD_Initialization_Node

```
1  InternalEvent(PD_INITIALIZATION_NODE_ON);
```

Listing 7.6: State transition for PD_Initialization_Node

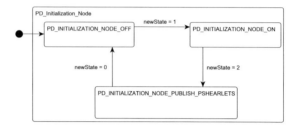

Figure 7.9: State machine for PD_Initialization_Node.

PD_Initialization_Node   The PD_Initialization_Node performs the calculation of the shear-lets and their transmission to other nodes in the system. Figure 7.9 shows the state machine contained in this node. On execution of the launch file, the init state PD_INITIALIZATION_NODE_OFF is entered. No tasks are executed in this state. The transition to the state PD_INITIALIZATION_NODE_ON takes place when the service PD_Initialization_Node_Init_srv is called. In this state, the calculation of the shearlets is performed. If the calculation is finished successfully, the transition to the state PD_INITIALIZATION_NODE_PUBLISH_PSHEARLETS is performed. Here, the calculated shearlet parameters are written to a topic. Subsequently, the state PD_INITIALIZATION_NODE_OFF is entered. The transition to the end state is performed on termination of the program and can be reached from any other state.

PD_Control_Node   This node is responsible for the flow control in our system. Figure 7.10 shows the state machine contained in the node. On execution of the launch file, the init state PD_CONTROL_NODE_OFF is entered. No tasks are executed in this state. No further event is needed for the state transition to PD_CONTROL_NODE_INIT_OPTS. It occurs directly after the previous state transition. In this state, all system parameters are loaded and initialized. After completion of the initialization, the shearlet filters are calculated in the state PD_CONTROL_NODE_INIT_CLASSIFIER. Here, the service of the PD_Initialization_Node is called to compute the shearlets. Subsequently, the classifier is initialized. In this state, it is identified if there is a classifier file already available or not. If there is no classifier file available, the state PD_CONTROL_NODE_TRAINING is entered. In case there is already a classifier available, we enter the state PD_CONTROL_NODE_DETECTION. The state PD_CONTROL_NODE_TRAINING calls a service from the PD_Training_Node, leading to training and saving of a classifier. Next, we have the state transition to PD_CONTROL_NODE_DETECTION. Here, a service from the PD_Detection_Node is called to start the detection. If the detection has been started successfully, the state PD_CONTROL_NODE_ON is entered. In this state, the node sleeps and performs no tasks. The transition to the end state is performed on termination of the program and can be reached from any other state.

PD_Training_Node   The PD_Training_Node is responsible for the training of the classifier. In Figure 7.11, we have its state machine. On execution of the launch file, the init state PD_TRAINING_NODE_OFF is entered. No tasks are executed in this state. The state transition to PD_TRAINING_NODE_ON occurs when the service PD_Training_Node_Train_srv is called. In this state, the training of the classifier is performed. When the training is completed, the state PD_TRAINING_NODE_OFF is entered. The transition to the end state is performed on termination of the program and can be reached from any other state.

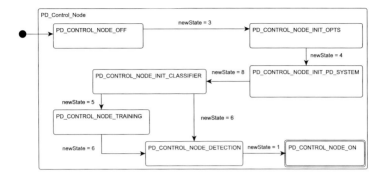

Figure 7.10: State machine for PD_Control_Node.

Figure 7.11: State machine for PD_Training_Node.

PD_Calculation_Node   The PD_Calculation_Node is the node where all fundamental calculations are performed. Figure 7.12 shows the state machine contained in the node. On execution of the launch file, the init state PD_CALCULATION_NODE_OFF is entered. No tasks are executed in this state. Subsequently, the state transition to PD_CALCULATION_NODE_ON occurs. In this state, all action-servers responsible for the corresponding calculations are started. The transition to the end state is performed on termination of the program and can be reached from any other state.

PD_Detection_Node   This node is responsible for the pedestrian detection on given input images. Figure 7.13 shows the state machine contained in the node. On execution of the launch file, the init state PD_DETECTION_NODE_OFF is entered. No tasks are executed in this state. The state transition to PD_DETECTION_INIT_OPTS occurs when the service PD_Detection_Node_Detect_srv is called. Here, all required parameters for the detection are initialized. When the initialization is completed, the state PD_DETECTION_NODE_STARTIO is entered. In this state, the service of the PD_InputOutput_Node is called to start loading the input images. When this is completed successfully, the state transition to PD_DETECTION_NODE_ON occurs. Actually in this state, we

Figure 7.12: State machine for PD_Calculation_Node.

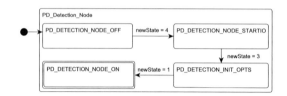

Figure 7.13: State machine for PD_Detection_Node.

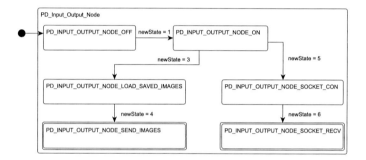

Figure 7.14: State machine for PD_Input_Output_Node.

have the pedestrian detection. The transition to the end state is performed on termination of the program and can be reached from any other state.

PD_Input_Output_Node This node is the interface to the input images as well as the visualization devices. Figure 7.14 shows the state machine contained in the node. On execution of the launch file, the init state PD_INPUT_OUTPUT_NODE_OFF is entered. No tasks are executed in this state. The state transition to PD_INPUT_OUTPUT_NODE_ON occurs when the service PD_Input_Output_Node_Start_srv is called. In this state, it is decided if input images from the hard disk are loaded or if images are received via TCP/IP. This decision is made on the value of the flag mode, which is set by the initial configuration. If this flag has the value IO_MODE_SAVED_IMAGES, the state PD_INPUT_OUTPUT_NODE_LOAD_SAVED_IMAGES is entered. Here, the saved input images are loaded and subsequently written on a topic in the state PD_INPUT_OUTPUT_NODE_SEND_IMAGES. If the flag mode has the value IO_MODE_SOCKET_CONNECTION, the state transition to PD_INPUT_OUTPUT_NODE_SOCKET_CON occurs. Here, an incoming TCP/IP connection is awaited. In case of a successful connection, the state PD_INPUT_OUTPUT_NODE_SOCKET_RECV is entered. In this state, the TCP/IP client receives images on which a pedestrian detection is performed. The result is then sent back to the TCP/IP client. The transition to the end state is performed on termination of the program and can be reached from any other state.

PD_Visualization_Node The PD_Visualization_Node displays the detection results. Figure 7.15 shows the state machine contained in the node. On execution of the launch file, the init

Figure 7.15: State machine for PD_Visualization_Node.

state PD_VISUALIZATION_NODE_OFF is entered. No tasks are executed in this state. Subsequently, the state PD_VISUALIZATION_NODE_ON is entered in which input data is awaited. If there is data available, it is displayed. The transition to the end state is performed on termination of the program and can be reached from any other state.

7.3.3 Partitioning

Via a CUDA interface, the NVIDIA Jetson TK1 provides the ability to use a GPU for calculation purposes. For this, NVIDIA has developed a specific hardware architecture. The following description of it is based on [88]. Figure 7.16 illustrates the connection between CPU and GPU. As can be seen, the CPU accesses GPU kernel functions via wrapper functions. These are functions which are distributed by the scheduler on a scalable array of multithreaded *Streaming Multiprocessors (SMs)*. A SM is comparable to a CPU in a sense that it has a register, an instruction memory, multiple cores and a shared memory such that threads can exchange data. It is designed to execute hundreds of threads in parallel. This ability is achieved by a *Single-Instruction Multiple Thread (SIMT)* architecture in which a scheduler clusters 32 threads. These threads are jointly generated, controlled and can execute the same instruction in one step. This clustering procedure is also called *warp*. Threads can be jointly started at the same memory address, however they have their own counters and registers. Optimally, each thread of a warp executes the same instruction to achieve the highest performance. The abovementioned kernels work on logical data blocks, which the scheduler partitions on the SMs. On the other hand, SMs partition these blocks into different warps. For more details on the NVIDIA hardware architecture, see [88]. As can be seen in Figure 7.16, our overall system contains kernel functions of the pedestrian detection system, OpenCV and NVIDIA. Kernel functions from the pedestrian detection system are GPU functions developed in this thesis. Whereas the other kernel functions are GPU functions provided by the OpenCV library and NVIDIA, respectively.

Figure 7.17 shows the CPU and GPU interfaces by the aid of component diagrams. For each of the illustrated wrapper and kernel functions, we have homogeneous CPU functions. By the initial configuration of the system, it can be chosen if the system shall be run with GPU support or as a CPU only variant.

7.4 Implementation

In this section, we will describe the crucial topics concerning implementation of the system. It is of major importance to achieve balance between memory consumption and computational power. The underlying Matlab algorithm is implemented only in regards to detection quality. The interaction of detection quality, memory consumption and computational power is regarded only as a subsidiary. Therefore, the Matlab implementation can not be transferred directly to C++ but only output-driven.

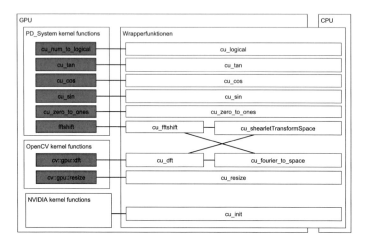

Figure 7.16: Connection of CPU and GPU via wrapper functions.

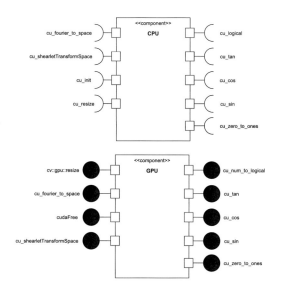

Figure 7.17: CPU and GPU interfaces.

```
1   <launch>
2   <rosparam param="opts/nWeak">[32,128,1024,2048] </rosparam>
3   <rosparam param="opts/pLoad/squarify">[3.0,0.41] </rosparam>
4   <rosparam param="opts/pPyramid/minDs">[100,41] </rosparam>
5
6   <node
7       pkg="embedded_pd"
8       type="pd_detection_node"
9       name="pd_detection"
10      output="screen"
11      launch-prefix="xterm -e"
12  />
13  </ launch >
```

Listing 7.7: Launch file example.

```
1   void simple(int n, float *a, float *b)
2   {
3       int i;
4
5       # pragma omp parallel for
6       for (i =1; i<n; i++){} /* i is private by default */
7           b[i] = (a[i] + a[i -1]) / 2.0;
8   }
```

Listing 7.8: Example of an OpenMP directive.

As described earlier, the system is started by the execution of a launch file. Such a ROS launch file contains init and configuration information which can be processed by the tool `roslaunch`. Listing 7.7 shows an example of a launch file for three parameters and one node.

7.4.1 Optimization

To gain performance boosts, our system is extended in terms of multithreading. For this purpose, OpenMP is used to generate and maintain threads. OpenMP is an acronym for Open specifications for Multi Processing. It holds compiler directives, library functions and environment variables and provides bindings for the programming languages C, C++ and FORTRAN [108]. These languages are extended by constructs for execution of programs with different data by multiple threads, for distribution of tasks to multiple threads, for synchronization of threads and for declaration of shared and private variables of threads. Parts of a serial program can be parallelized by only few directives. It gives one the ability of a parallel processing on multiple processors, such that runtime can be reduced significantly. Listing 7.8 shows an example of the simplicity using OpenMP. Here, a `for` loop is distributed on the available cores.

This example is a simple construct, which can be extended by a diverse number of complex directives. In Listing 7.9 we exemplarily show the implementation of the shearlet transform in space domain. The contained `for` loop is parallelized by OpenMP. The difference to the `for` loop in Listing 7.8 is that each thread requires data which it reads and writes. To guarantee a secure access to the data, OpenMP provides different clauses. In this example, we have the clauses `shared`, `private`, `firstprivate` and `ordered`. The clauses `shared` and `private` specify if the data can be accessed by all threads or only by one. The clause `firstprivate` is used to initialize private variables with the value of the master thread. The clause `ordered` references to the scheduling of the threads and their access on the `shared` object.

```
1   void pd::shearletTransformSpace(Mat & I, vector<Mat>& Psi, vector<Mat>& C, int
        flag_fourier)
2   {
3       Mat w,conv;
4       int nShearlets = Psi.size();
5       C.reserve(nShearlets);
6
7       # pragma omp parallel for firstprivate (flag_fourier) private (w,conv) shared
            (C ,Psi ,I) ordered schedule (dynamic)
8       for (int i = 0; i < nShearlets ; i++)
9       {
10          if (flag_fourier)
11          {
12              w = Psi.at(i);
13              fftshift(w);
14              idft(w, DFT_COMPLEX_OUTPUT);
15              fftshift(w);
16          }
17          else
18          {
19              w = Psi.at(i);
20          }
21          conv2(I, w, CONVOLUTION_SAME , conv);
22          conv = abs(conv);
23
24          # pragma omp ordered
25          C.push_back(conv.clone());
26      }
27  }
```

Listing 7.9: Implementation of the shearlet transform in space domain using OpenMP directives.

```
1  #!/bin/sh
2  # Set CPU to full performance on NVIDIA Jetson TK1 Development Kit
3  if [ $(id -u) != 0 ]; then
4      echo " Need root permissions "
5      echo "$ sudo "$0""
6      exit
7  fi
8
9  cd /sys/devices/system/cpu
10 grep -H . cpuquiet/tegra_cpuquiet/enable
11 grep . cpu?/online
12 grep . cpu?/cpufreq/scaling_governor
13
14 echo 0 >cpuquiet/tegra_cpuquiet/enable
15 for i in cpu?; do
16     read online <$i/online
17     [ "$online" = 1 ] && continue
18     echo 1 >$i/online
19 done
20 for i in cpu?; do
21     echo performance >$i/cpufreq/scaling_governor
22 done
23
24 grep -H . cpuquiet/tegra_cpuquiet/enable
25 grep . cpu?/online
26 grep . cpu?/cpufreq/scaling_governor
```

Listing 7.10: NVIDIA Jetson TK1 CPU optimization.

Besides the optimization of the calculation processes, we also have to consider to optimize the embedded target itself. The Tegra K1 chipset of the NVIDIA Jetson TK1 development kit has been developed for mobile applications. It contains several systems which maintain the hardware. These systems can either accelerate or decelerate the hardware. On delivery, the CPU and the GPU are not configured to the maximum performance. Listings 7.10 and 7.11 show the shell scripts which are used to set the high-performance configurations of CPU and GPU.

The NVIDIA Jetson TK1 has 2GB of RAM, of which 50% are consumed by the operation system. Therefore, only 1GB of RAM is available for calculation. Unfortunately, the training process requires 6GB. Therefore, the embedded target does not perform a training process and receives a classifier trained on the PC.

Concerning runtime performance, an efficient application of the ROS architecture results in significant gains. We illustrate this on the example of the training process. We consider that $1,000$ images of the training data set shall be loaded to train the classifier. A sequential procedure of the training process is shown in 7.18. Except the calculations of the PD_Calculation_Node, which are distributed to multiple CPU cores, the complete process runs only on one CPU core.

Intuitively, an optimization of this procedure is beneficial. For this purpose, 4 action-servers with 4 action-clients are set up, where each one processes a portion of the training images. This procedure is shown in 7.19. With this solution, all CPU cores perform calculations with maximum performance, such that the runtime of the training phase can be reduced significantly.

```
1   #!/bin/sh
2   # Set CPU to full performance on NVIDIA Jetson TK1 Development Kit
3   if [ $(id -u) != 0 ]; then
4       echo "Need root permissions"
5       echo "$ sudo "$0""
6       exit
7   fi
8
9   cd /sys/kernel/debug/clock
10
11  for clock in gbus emc; do
12      grep -H . $clock/rate
13      read rates <$clock/possible_rates
14      for rate in $rates; do
15          [ "$rate" = '(kHz)' ] && break
16          max_rate = $rate
17      done
18      echo ${ max_rate }000 >override.$clock/rate
19      echo 1 >override.$clock/state
20      grep . override.$clock/rate override.$clock/state
21  done
```

Listing 7.11: NVIDIA Jetson TK1 GPU optimization.

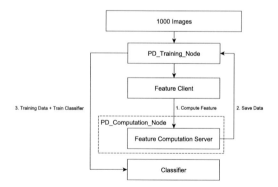

Figure 7.18: Sequential process of the training phase.

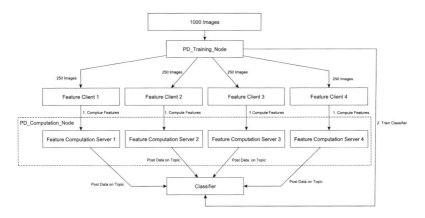

Figure 7.19: Multiaction structure of the training process.

7.4.2   HW/SW Codesign

The support of the GPU and its usage as a coprocessor can result in significant increase of computational performance. In this section, we analyze the source code and identify possible processes in which the parallel computing using the GPU can be beneficial. One of these processes is the computation of the shearlet transform from a color image, for example in LUV color space. Here, each color channel is convolved with all shearlet filters. Thus, if we have for example 18 shearlets it results in 54 convolutions. Since this calculation consumes a lot of runtime, it is a suitable candidate for using the GPU as coprocessor. Figure 7.20 illustrates the concept of how the GPU can be used in this case. The GPU is intended to perform the convolutions with one color channel while the CPU takes care of the remaining color channels.

Surprisingly, the anticipated speed up does not take place. Instead the calculation time is even decelerated. The main reason for this fact is that data has to be transferred from CPU to GPU before calculation and vice versa after it. This context switch results in a faster runtime of the CPU only solution. Therefore, a deeper analysis is needed how the GPU can be used as a coprocessor to spare runtime. For this purpose, one can use genetic algorithms [43] for optimization. Genetic algorithms are based on the concepts of natural selection and natural genetics and used as search algorithms. An evolutionary process is executed in which selection, recombination and mutation are performed. Through such procedures, random combination of components are assembled and selected by best results. Thus, one gets an optimal assembly of software and hardware components. This procedure is used for the analysis of the GPU improvement abilities. As a result, no tested permutation with the GPU yields in a performance improvement. The system performs better with the CPU only solution. Therefore, the issue of HW/SW codesign fades from the spotlight concerning runtime optimization. We leave a deeper analysis and a more sophisticated development concerning GPU usage as an open topic for future works. Especially, embedded targets with more powerful GPU performance should be considered since it is immensely increasing at the moment.

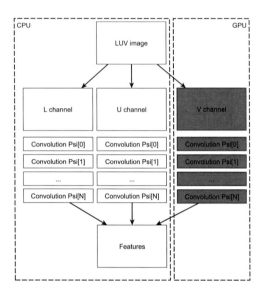

Figure 7.20: Calculation process with GPU as coprocessor.

7.5   Conclusion

In this chapter, we described the implementation of the pedestrian detection algorithm for an embedded target. As a result, we have a version of the detection algorithm shearFtrs-v1 runable with 10fps on the NVIDIA Jetson TK1 target. The architectural basis is the Robot Operating System, which leads to a high modularity and extensibility of our system. By the encapsulation of the detection algorithm on multiple nodes, the system can be distributed to diverse platforms. Thus, resources for calculation, control and visualization processes can be separated. By the ROS parameter server, our system is simply parameterized and has user-friendly configuration interfaces. For this purpose, the system can be configurated and started externally by a launch file.

Concerning runtime optimization, an efficient usage of the ROS architecture and the application of multithreading yield major performance boosts. Contrary to expectations, the usage of a GPU as coprocessor does not yield any performance improvements. A deeper analysis and a more sophisticated development concerning GPU usage is left as a perspective for future works, especially by using recently released embedded targets, for which the GPU performance has dramatically increases at present.

"Every new beginning comes from some other beginning's end."

Seneca

8

Conclusions and Perspectives

In this thesis, we have addressed the application of shearlets for image feature extraction both from a theoretical and a practical point of view. We paid special attention to the usage in pedestrian detection algorithms, since these are currently a key topic in the area of computer vision and artificial intelligence. In that sense, we analyzed the applicability of shearlets to improve the detection quality of current state-of-the-art approaches. These approaches split up in two main categories. First, so-called classical pedestrian detection algorithms computing hand-crafted image features and subsequent application of a machine-learning procedure. Second, detection algorithms based on Convolutional Neural Networks. Currently, the best performing algorithms belong to the second category. In order to have a complete analysis, we studied the improvement ability of hand-crafted image features based on shearlets as well as the shearlet integration into CNNs. To round up, we examined the deployability of a pedestrian detection algorithm which requires the computation of the shearlet transform on an embedded system, e.g. an electronic control unit. To that end, we evaluated if such an algorithm can be run on a reasonably priced embedded target with a practical usable frame rate.

In Chapter 2, we reviewed the existing shearlet frameworks. We first concentrated on providing some theoretical key results from the recent literature on shearlet frameworks. Next, we turned to the analysis of current realizations, especially in regards to their usability for image feature extraction. We found that most frameworks concentrate on fulfilling tight frame properties. To achieve that aim, band-limited shearlet mother functions are commonly used. This leads to shearlets with infinite support in time domain and thus to improvable localization of edges in images. Current frameworks using compactly supported shearlets, such as ShearLab 3D [77], produce shearlets with support sizes still having room for improvement concerning precision in localization of edges.

Given our considerations about the improvability of edge detection using shearlets we set up our own shearlet design in Chapter 3. We applied compactly supported, separable mother functions which are point-symmetric in the first and axis-symmetric in the second component. We call shearlets based on these mother functions *local precision shearlets*. The corresponding shearlet system is set up such that we have a high amount of flexibility concerning the number of shearlets per scale. This property is essential for controlling the dimension of the image feature space used for pedestrian detection. We showed that our shearlet system forms a frame for $L^2 (\mathbb{R}^2)$ and provided the conditions that have to be fulfilled for this statement.

In Chapter 4, we dealt with the properties of local precision shearlets concerning edge detection from a theoretical point of view. We showed that, using our shearlets, edge points in $\mathbb{R}^2$ and their type can be characterized by the limits of the shearlet transform for decreasing scales. Furthermore, we showed which property the shearlet mother function has to fulfill and gave corresponding examples.

Turning to the practical side, we described the pedestrian detection using shearlet features in Chapter 5. We defined two types of image features based on the shearlet transform, namely the shearlet magnitude and shearlet histogram features. Moreover, we defined a filterbank based on shearlets for an intermediate filtering layer between feature computation and classification. Based on our experimental results on the Caltech data set [30], our features turned out to be very informative. We showed that they are able to outperform all other hand-crafted features currently used.

Recently, various pedestrian detection algorithms using CNN models were set up [11, 12, 32, 84, 117, 127]. In CNN models, features are not hand-crafted any longer but learned during the training process. Current results on the Caltech pedestrian detection benchmark data set [30] show better detection results of approaches using CNN models. Due to the current prevalence of CNN approaches for pedestrian detection, we studied the possibility to integrate shearlets in CNN models in Chapter 6. We used shearlet filters at early convolution layers of a CNN instead of learned ones. Our reasoning for this approach is that the first layers of a CNN perform an edge detection and that shearlets theoretically represent optimal filters for this topic. In order to harmonize the theory with approaches showing the best results in practice, only a portion of the first layer filters is initialized by shearlets. We initialize and train the remaining filters as in the corresponding state-of-the-art procedures. In our experiments concerning pedestrian classification, we found that this approach applied to the network architecture of AlexNet [72] yields improvements of classification results compared to the original AlexNet trained on the same data. In our experiments concerning the detection of pedestrians, we found that we need deeper architectures like the VGG16 [113] in order to achieve results in the range of currently best performing algorithms. We achieved improvements of the detection quality by a shearlet initialization of the VGG16 model [113] trained on the CIFAR-100 data set. However, we could not obtain the results of Brazil et al. [9], who applied the VGG16 pre-trained on the extensive ImageNet data. We found this result as an indication for the immense power of data in deep learning algorithms and concluded that shearlets cannot ease the need of a pre-training with huge and comprehensive data sets. We left the evaluation of the shearlet integration with a pre-training on an extensive data set such as ImageNet as a perspective for further research.

In Chapter 7, we described how the shearlet transform and the base algorithm presented in Chapter 5 can be ported to an embedded target. We re-implemented the algorithm in C++ and used the NVIDIA Jetson TK1 developer kit [98] as an example for a suitable embedded platform. Given this reasonably priced target, we were able to run the detection algorithm with a useful frame rate of 10fps. A next step would be to further boost the runtime by an efficient use of the GPU. Especially under usage of recently released embedded targets, for which the GPU performance has dramatically increased in the last few years.

In summary, we found that the framework of shearlets provides the ability to improve both prevalent types of pedestrian detection. Classical algorithms using hand-crafted image features as well as approaches utilizing CNN models. In order to achieve that goal we first needed to carefully design shearlets such that they have compact support in time domain and can be digitalized to filters with small pixel size. Second, we needed to build a shearlet system that is more flexible in terms of shearlets per scale. In that way, we were able to better control the feature space size of our hand-crafted features as well as the number of shearlets integrated into a CNN. Moreover, the corresponding sampling strategy of the shearing parameter provided the possibility to apply an intermediate filterbank layer with shearlet filters on top of gradient histogram features. This led to the currently best performing hand-crafted image features in the Caltech Pedestrian Detection Benchmark [30]. Moreover, we provided an embedded realization of the shearlet transform and a detection algorithm based on it with a practical useful frame rate by a mindful system and software architecture as well as optimized implementation.

This means that we needed to carefully adapt the theory on shearlets according to the requirements of the practical application. On the other hand, a mindful theoretical treatment of the shearlet framework enabled the improvement of state-of-the-art procedures in practice. In other words, only by a careful consideration of both sides, our merge of theory and practice led to success.

Spaces of Functions

In this chapter, we briefly describe the spaces of functions used in this thesis. The description is based on the definitions in [1]. In the following, we will consider Ω as an open subset of $\mathbb{R}^d$.

A basic but important space is the one of continuous functions $C\left(\Omega\right)$. In addition, $C^m\left(\Omega\right)$ denotes the space of m-times *continuously differentiable functions*. For a differentiable function f defined on Ω and a *multi-index* $\alpha = (\alpha_1, \ldots, \alpha_d) \in \mathbb{Z}_+^d$, we write $|\alpha| = \sum_{j=1}^d \alpha_j$, $x^\alpha = x_1^{\alpha_1} \cdots x_d^{\alpha_d}$ and

$$D^\alpha = \frac{\partial^{\alpha_1}}{\partial x_1^{\alpha_1}} \cdots \frac{\partial^{\alpha_d}}{\partial x_1^{\alpha_d}}$$

for the *partial derivative* of it. $C^m\left(\Omega\right)$ consists of all functions f defined on Ω that are continuous and for which their partial derivatives $D^\alpha f$ of orders $|\alpha| \leq m$ are continuous on Ω. For the infinitely many-times continuously differentiable functions, we set $C^\infty\left(\Omega\right) = \bigcap_{m=0}^\infty C^m\left(\Omega\right)$. The subspaces of $C\left(\Omega\right)$, $C^m\left(\Omega\right)$ and $C^\infty\left(\Omega\right)$, consisting of all functions that are compactly supported in Ω, are denoted by $C_0\left(\Omega\right)$, $C_0^m\left(\Omega\right)$ and $C_0^\infty\left(\Omega\right)$.

As Kutyniok and Labate [75] describe, the space of *square-integrable functions* on $\mathbb{R}^d$, denoted by $L^2(\mathbb{R}^d)$, is the standard model for signals in $\mathbb{R}^d$. In general, for $1 \leq p < \infty$ the space $L^p\left(\Omega\right)$ consists of all functions f defined on Ω that fulfill

$$\int_\Omega |f\left(x\right)|^p \, \mathrm{d}x < \infty.$$

The space $L^p\left(\Omega\right)$ is equipped with the L^p-Norm $\|\cdot\|_p$ defined by

$$\|f\|_p = \left(\int_\Omega |f\left(x\right)|^p \, \mathrm{d}x\right)^{1/p},$$

while the corresponding inner product is given by

$$\langle f, g \rangle = \int_\Omega f\left(x\right) \overline{g\left(x\right)} \mathrm{d}x.$$

Next, we take care of the notion of an *approximate identity* in L^p spaces. A sequence of functions $(\rho_n)_{n=1}^\infty$ is called an approximate identity if $\rho_n \geq 0$, $\int_{\mathbb{R}^d} \rho_n\left(x\right) \mathrm{d}x = 1$ and for every $f \in L^1(\mathbb{R}^d)$, we have $\lim_{n\to\infty} \|\rho_n * f - f\|_1 = 0$.

Finally, we state the main definitions concerning *Sobolev spaces*. At first, we need the definition of *weak derivatives*. Let $f, g_\alpha \in L^1_{loc}(\Omega)$, i.e. $f, g_\alpha \in L^1(\Omega)$ for every open subset $U \subset \Omega$. We say that g_α is the *weak derivative* of f if

$$\int_\Omega f(x) D^\alpha \phi(x) \, dx = (-1)^{|\alpha|} \int_\Omega g(x) \phi(x) \, dx$$

for every $\phi \in C_0^\infty(\Omega)$. For $m \in \mathbb{N}^+$, $1 \le p < \infty$ and $f \in L^p(\mathbb{R}^d)$, we define the *Sobolev norm* $\|\cdot\|_{m,p}$ as follows:

$$\|f\|_{m,p} = \left(\sum_{0 \le |\alpha| \le m} \|D^\alpha f\|_p^p \right)^{1/p}.$$

We define the vector space

$$H^{m,p}(\Omega) = \left\{ f \in L^p(\Omega) : D^\alpha f \in L^p(\mathbb{R}^2), 0 \le |\alpha| \le m \right\},$$

where $D^\alpha u$ is the weak derivative. Equipped with the norm $\|\cdot\|_{m,p}$, $H^{m,p}(\Omega)$ is called *Sobolev space* over Ω.

B

Basic Fourier Analysis

This chapter contains the basic definitions and results concerning Fourier analysis used in this thesis. The description is based on [45]. In general, Fourier analysis treats the analysis of functions concerning their Fourier transform. The Fourier transform of a signal $f \in L^1(\mathbb{R}^d)$ is defined by

$$\hat{f}(x) = \int_{\mathbb{R}^d} f(x) e^{-2\pi i \langle x, \xi \rangle} \mathrm{d}x,$$

where $\langle x, \xi \rangle = \sum_{i=1}^d x_i \xi_i$ is the inner product in $\mathbb{R}^d$. In some cases we will also write $\mathcal{F}(f)$ instead of $\hat{f}$. A fundamental result concerning Fourier analysis is *Plancharel's theorem* which is given as follows.

Theorem B.1. *For $f \in L^1(\mathbb{R}^d) \cap L^2(\mathbb{R}^d)$ we have*

$$\|f\|_2 = \|\hat{f}\|_2.$$

Consequently, $\mathcal{F}$ satisfies the Plancharel formula for $f, g \in L^2(\mathbb{R}^d)$, which is given by

$$\langle f, g \rangle = \langle \hat{f}, \hat{g} \rangle.$$

An interpretation of Plancharel's theorem is that the Fourier transform preserves the energy of a signal f.

In the following, we describe how the Fourier transform behaves for some basic operations on functions on which the transform is applied. For $f \in L^1(\mathbb{R}^d)$ and $x, \xi \in \mathbb{R}^d$ we define the *translation* by x as

$$T_x f(t) := f(t - x)$$

and the *modulation* by ξ as

$$M_\xi f(t) := e^{2\pi i \langle \xi, t \rangle} f(t).$$

For a translated or modulated signal $f \in L^1(\mathbb{R}^d)$, we have

$$\widehat{T_x f} = M_{-x} \hat{f}$$

and
$$\widehat{M_\xi f} = T_\xi \hat{f}.$$
Due to the last formula, modulations are also called *frequency shifts*. Similarly, translations are also called *time shifts*. A combination of the last two formulas yields
$$\widehat{T_x M_\xi f} = M_{-x} T_\xi \hat{f} = e^{-2\pi i \langle x, \xi \rangle} T_\xi M_{-x} \hat{f}.$$
A main functional operation in this thesis is the convolution of two functions f, $g \in L^1(\mathbb{R}^d)$. It is defined by
$$(f * g)(x) = \int_{\mathbb{R}^d} f(y) g(x - y) \, dy$$
and satisfies
$$\|f * g\|_1 = \|f\|_1 \|g\|_1.$$
For the Fourier transform of the convolution of two functions we get
$$\widehat{(f * g)} = \hat{f} \cdot \hat{g}.$$

Another topic of interest is the Fourier transform of derivatives. As before, given a multi-index $\alpha = (\alpha_1, \ldots, \alpha_d) \in \mathbb{Z}_+^d$, we write $D^\alpha f$ for the partial derivative of f and $x^\alpha = x_1^{\alpha_1} \cdots x_d^{\alpha_d}$. For the Fourier transform of it, we get
$$\widehat{D^\alpha f} = (2\pi i \xi)^\alpha \hat{f}$$
and
$$((-\widehat{2\pi i x})^\alpha f) = D^\alpha \hat{f}.$$
Finally, we state the *inversion formula* for the Fourier transform.

Theorem B.2. *For $f \in L^1(\mathbb{R}^d)$ and its Fourier transform $\hat{f} \in L^1(\mathbb{R}^d)$, we have*
$$f(x) = \int_{\mathbb{R}^d} \hat{f}(\xi) e^{2\pi i \langle x, \xi \rangle} d\xi, \quad \text{for all } x \in \mathbb{R}^d.$$

An Alternative Approach on Shearlet Design

In the following, we design a local precision shearlet as alternative to the ones described in Section 4.2 that is appealing concerning boundary definition and discretization and mimics characteristics of spline shearlets to fulfill condition (4.5). Our method to design a well shapeable mother function is closely related to the construction of the famous *Meyer wavelet* [93], which is originally defined in frequency domain. Consequently they are used by Kutyniok et al. [80] as well as Häuser and Steidl [61] for their definition of shearlets in frequency domain. We adapt this idea to define shearlets in time domain. According to Häuser [60], Meyer wavelets can be characterized by freely selectable parameters of a center x_0, a starting point A and a support width d_0. Our approach to define the shearlet components ψ_1 and ψ_2 is mainly adapted from the description of Meyer-type wavelets by Häuser [60].

We start by using a function $v \colon \mathbb{R} \to \mathbb{R}$ with $v \in C^k$, $k \in \mathbb{N}$, and

$$v\left(x\right) = \begin{cases} 0 & \text{for } x \leq 0, \\ 1 - v\left(1 - x\right) & \text{for } 0 < x < 1, \\ 1 & \text{for } x \geq 1 \end{cases}$$

to define function $w_1 \colon \mathbb{R} \to \mathbb{R}$ by

$$w_1\left(x\right) = \begin{cases} \sin\left(\frac{\pi}{2} v\left(px - q\right)\right) & \text{for } A \leq x \leq x_0, \\ \cos\left(\frac{\pi}{2} v\left(rx - q\right)\right) & \text{for } x_0 < x \leq A + d_0\,, \\ 0 & \text{otherwise} \end{cases}$$

with A, x_0, $d_0 \in \mathbb{R}_0^+$ and parameters $p, q, r \in \mathbb{R}$. According to [60], with $v \in C^k$ it follows that also $w_1 \in C^k$. There are many possible choices to define an appropriate function v depending on the desired order of smoothness k. Following the proposal of Kutyniok et al. [80], we use $v \in C^1$ given by

$$v\left(x\right) = \begin{cases} 0 & \text{for } x < 0, \\ 2x^2 & \text{for } 0 \leq x \leq \frac{1}{2}, \\ 1 - 2\left(1 - x\right)^2 & \text{for } \frac{1}{2} < x \leq 1, \\ 1 & \text{for } x > 1. \end{cases}$$

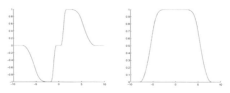

(a) Wavelet ψ_1 and bump function ψ_2.

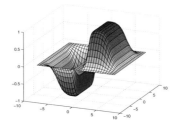

(b) Shapeable shearlet $\psi(x) = \psi_1(x_1)\psi_2(x_2)$.

Figure C.1: Example of a shapeable shearlet and its components for $a_0 = 1/4$ and $x_0 = 2$.

According to Häuser [60], the parameters for w_1 are given by $p = 1/x_0(1-a_0)$, $q = a_0/(1-a_0)$, $r = a_0/x_0(1-a_0)$ with $0 < a_0 \leq 1$. It follows that $A = x_0 a_0$ as well as $A + d_0 = x_0/a_0$ and therefore

$$w_1(x) = \begin{cases} \sin\left(\frac{\pi}{2}v\left(px - q\right)\right) & \text{for } x_0 a_0 \leq x \leq x_0, \\ \cos\left(\frac{\pi}{2}v\left(rx - q\right)\right) & \text{for } x_0 < x \leq \frac{x_0}{a_0}, \\ 0 & \text{otherwise.} \end{cases}$$

Originally, Meyer wavelets are defined as even functions given v and w_1. To fulfill the condition on the wavelet component of a local precision shearlet, we make use of the sign function sgn. Let $\psi_1 \colon \mathbb{R} \to \mathbb{R}$ be defined by

$$\begin{aligned} \psi_1(x) &= \operatorname{sgn}(x)\, w_1(|x|) \\ &= \begin{cases} \operatorname{sgn}(x)\sin\left(\frac{\pi}{2}v\left(p\,|x| - q\right)\right) & \text{for } x_0 a_0 \leq |x| \leq x_0, \\ \operatorname{sgn}(x)\cos\left(\frac{\pi}{2}v\left(r\,|x| - q\right)\right) & \text{for } x_0 < |x| \leq \frac{x_0}{a_0}, \\ 0 & \text{otherwise.} \end{cases} \end{aligned} \tag{C.1}$$

For the bump component ψ_2, we use the scaling function of the Meyer wavelet which is given by

$$w_2(x) = \begin{cases} 1 & \text{for } 0 \leq |x| \leq a_0 x_0, \\ \cos\left(\frac{\pi}{2}v\left(px - q\right)\right) & \text{for } a_0 x_0 \leq |x| \leq x_0, \\ 0 & \text{otherwise.} \end{cases}$$

To ensure the symmetry of ψ in $\mathbb{R}^2$, we scale w_2 such that we get the same support boundary as for ψ_1. Consequently, we define $\psi_2 \colon \mathbb{R} \to \mathbb{R}$ by

$$\psi_2(x) = w_2\left(a_0\,|x|\right)$$

$$= \begin{cases} 1 & \text{for } 0 \le |x| \le x_0, \\ \cos\left(\frac{\pi}{2} v \left(r \, |x| - q\right)\right) & \text{for } x_0 \le |x| \le \frac{x_0}{a_0}, \\ 0 & \text{otherwise.} \end{cases} \tag{C.2}$$

Definition C.1. We call $\psi = \psi_1 \psi_2$ with ψ_1 and ψ_2 defined by (C.1) and (C.2) a *shapeable shearlet*.

Since ψ is continuous and compactly supported, we obtain $\psi \in L^2\left(\mathbb{R}^2\right)$. Figure C.1 shows an example of the components ψ_1 and ψ_2 and the resulting shearlet using parameters $a_0 = {}^1/{}_4$ and $x_0 = 2$. Therefore, we have a support boundary $b = 8$ for both components in this example.

D

Notation, Symbols and Abbreviations

In this chapter, we recall some standard notation which was not defined in the main body or the previous parts of the Appendix. Furthermore, we will provide a list with symbols and abbreviations used in this thesis.

D.1   Standard Notation

For $f, g \colon \mathbb{R} \to \mathbb{R}$ and $a \in \mathbb{R}$, we write $f(x) = \mathcal{O}(g(x))$ for $x \to a$ if and only if there exist positive numbers δ and C such that

$$|f(x)| \leq C |g(x)|, \quad \text{when } 0 < |x - a| < \delta.$$

For $D \subset \mathbb{R}^d$, the *characteristic function* $\chi_D \colon \mathbb{R}^d \to \{0, 1\}$ is defined by

$$\chi_D(x) = \begin{cases} 1 & \text{for } x \in D, \\ 0 & \text{otherwise.} \end{cases}$$

Concerning practical application of the convolution defined in Appendix B, we state the definition of the *discrete convolution* of two sequences. Since we deal with images in this thesis, we consider the sequences to be of finite length. Let $f(n) = f(0), \ldots, f(M-1)$ and $g(n) = g(0), \ldots, g(N-1)$ be sequences in $\mathbb{R}^d$ of length $M, N \in \mathbb{N}$. Then, the discrete convolution of f and g is defined by

$$(f * g)(n) = \sum_{k=\max(0,n)}^{\min(n-M+1,N-1)} f(n) g(n-k).$$

D.2   Symbols

Sets, Spaces and Norms

$\mathbb{C}$	Set of all complex numbers
$\mathbb{N}$	Set of natural numbers without 0.
$\mathbb{N}_0$	Set of natural numbers with 0, $\mathbb{N}_0 = \mathbb{N} \cup \{0\}$
$\mathbb{R}$	Set of all real numbers
$\mathbb{R}_+$	Set of all real positive numbers without 0
$\mathbb{Z}$	Set of integers
$\mathbb{Z}_+$	Set of positive integers without 0
C^k	Space of functions $f \colon \mathbb{R} \to \mathbb{R}$ which are k-times continuously differentiable
$\mathcal{H}$	Hilbert space
$H^s(\Omega)$	Sobolev space of order s
$L^p(\Omega)$	Space of all measurable functions $f \colon \Omega \to \mathbb{C}$ such that $\int_\Omega |f(x)|^p \mathrm{d}x < \infty$ for $1 \le p < \infty$
$\mathcal{R}_\theta$	Set of rotations of angles $\theta = 2k\pi/K$ for $0 \le k < K \in \mathbb{N}$
$\|\cdot\|_X$	Norm of space X
$\|\cdot\|_\infty$	Supremum norm
$\langle \cdot, \cdot \rangle$	Scalar product of L^2

Operators and Functions

$\mathcal{A}$	Activation function
D^α	Partial derivative operator for multi-index $\alpha = (\alpha_1, \dots, \alpha_d) \in \mathbb{Z}_+^d$
$\frac{\partial}{\partial x_i}$	Partial derivative operator in x_i direction
∇	Gradient operator
$\mathcal{F}$	Fourier transform operator
$\hat{f}$	Fourier transform of function f, i.e. $\hat{f} = \mathcal{F}f$
$f * g$	Convolution of two functions f and g
M_ξ	Modulation operator (modulation by ξ)
$\mathcal{I}$	Identity operator
$\psi_{j,\theta}$	2-dimensional wavelet rotated by angle θ
$\psi_{j,\beta,\bar{j}}$	Roto-translation wavelet
S	Frame operator
$\mathcal{S}_p$	Windowed scattering transform for path p
$\tilde{\mathcal{S}}_p$	Scattering transform for path p
T_x	Translation operator (translation by x)
χ_D	Characteristic function

Shearlet Systems and Transforms

$A_{a,\alpha}$	Scaling matrix
A_a	Parabolic scaling matrix
b_1, b_2	Boundaries of a local precision shearlet
c	Sampling constant

$c_\psi^-, c_\psi^+, c_\psi$	Admissibility constants of a shearlet ψ
$\mathcal{C}_i$	Frequency domain cones, $i = 1, \dots, 4$
Γ_I	Irregular shearlet parameters
Γ_R	Regular shearlet parameters
$\Gamma, \tilde{\Gamma}$	Irregular one-adapted shearlet parameters
$\mathcal{LPSH}$	Local precision shearlet system
$LPST_{\phi,\psi,\tilde{\psi}}$	Continuous local precision shearlet transform
$\mathcal{LPST}_{\phi,\psi,\tilde{\psi}}$	Discrete local precision shearlet transform
$\mathcal{R}$	Low frequency region
S_s	Shearing matrix
$\mathbb{S}_L$	Parameter set for continuous local precision shearlet transform
$SH(\psi)$	Continuous shearlet system
$\mathcal{SH}_\psi$	Shearlet transform
$SH(\psi, \Gamma_I)$	Irregular shearlet system
$SH(\psi, \Gamma_R)$	Regular shearlet system
$SH(\phi, \psi, \tilde{\psi}, \Gamma, \tilde{\Gamma})$	Irregular cone-adapted shearlet system
$SH(\phi, \psi, \tilde{\psi}, c)$	Regular cone-adapted shearlet system
$\Phi, \Psi, \tilde{\Psi}$	Components of a cone-adapted shearlet system
ϕ_m	Low-frequency component of a cone-adapted shearlet system
$\psi_{j,k,m}$	Discrete shearlet
$\psi_{a,s,t}$	Continuous shearlet

Shearlet Features

$C_{j,k}$	Shearlet coefficient for scale j and shear k
ε	Normalization constant
$M \times N$	Image size
M_j	Shearlet magnitude for scale j
$\tilde{M}_j$	Normalized shearlet magnitude for scale j
O_j	Orientation of maximum shearlet coefficient at scale j
$\mathcal{P}_l$	Quadratic image patch
R	Normalization radius
$\mathcal{SH}_{j,k}$	Shearlet histogram feature for scale j and shear k
$\mathcal{SM}_j$	Shearlet magnitude feature for scale j
ζ	Patch size

Miscellaneous

$\vec{\alpha}(t)$	Arc-length parametrization of ∂R
$B_\epsilon(p)$	Ball of radius ϵ with center p
$\det(A)$	Determinant of matrix A
$\text{diag}(v)$	Diagonal matrix with entries v
Φ_Ω	Feature extractor based on module-sequence Ω
$\mathcal{G}$	Digital grid
γ_l	Gaussian initialized filter in layer l of a Convolutional Neural Network
H_l	Hidden layer l in an Artificial Neural Network
$\text{supp}\, f$	Support of function f, i.e. $\overline{\{x : f(x) \neq 0\}}$
$\vec{\kappa}(t^-), \vec{\kappa}(t^+)$	Curvature at point $\alpha(t)$ of ∂R

μ_p — Dirac response concerning the scattering transform of path p

$\left\langle \begin{matrix} n \\ k \end{matrix} \right\rangle$ — Eulerian number

$\vec{n}(t^-), \vec{n}(t^+)$ — Outer normals at point $\alpha(t)$ of ∂R

$\mathcal{O}(g)$ — Landau symbol

Ω — Module-sequence

$\mathfrak{S}_\rho$ — Domains with piecewise smooth boundary and curvature bounded by ρ

D.3   Abbreviations

ACF	Aggregated Channel Features
ADAS	Advanced Driver Assistance Systems
AEB	Autonomous Emergency Braking
ANN	Artificial Neural Network
CNN	Convolutional Neural Network
CPU	Central Processing Unit
CUDA	Compute Unified Device Architecture
eMMC	Embedded Multimedia Card
FFST	Fast Finite Shearlet Transform
FFT	Fast Fourier Transform
FPGA	Field Programmable Gate Array
FPPI	False Positives Per Image
GPU	Graphics Processing Unit
HDMI	High Definition Multimedia Interface
HOG	Histogram of Oriented Gradients
HW	Hardware
IFFT	Inverse Fast Fourier Transform
ILSVRC	ImageNet Large-Scale Visual Recognition Challenge
IP	Internet protocol
LAN	Local Area Network
LDCF	Local Decorrelation Channel Features
LPST	Local Precision Shearlet Transform
MRDS	Microsoft Robotics Developer Studio
NCAP	New Car Assessment Program
NMS	Non-Maximum Suppression
PC	Personal Computer
PD	Pedestrian Detection
R-CNN	Regions with CNN Features
ReLU	Rectified Linear Unit
ROC	Receiver Operating- Characteristic
RoI	Regions of Interest
ROS	Robot Operating System
RPN	Region Proposal Network
RSS	Really Simple Syndication
SATA	Serial AT Attachment
SCED	Shearlet Cascade Edge Detection
SDS-RCNN	Simultaneous Detection and Segmentation R-CNN
SI-CNN	Shearlet Initialized CNN

SIMT	Single-Instruction Multiple Thread
SM	Streaming Multiprocessor
SOC	System-on-a-Chip
SSD	Single Shot Multibox Detector
SVM	Support Vector Machines
SW	Software
TCP	Transmission Control Protocol
UDP	User Datagram Protocol
XML-RPC	Extensible Markup Language Remote Procedure Call

Bibliography

[1] Robert A. Adams and John J.F. Fournier. *Sobolev Spaces*. Pure and Applied Mathematics. Elsevier Science, 2003.

[2] Hamed Habibi Aghdam and Elnaz Jahani Heravi. *Guide to Convolutional Neural Networks*. Springer International Publishing, 1st edition, 2017.

[3] Jean-Pierre Antoine, Pierre Carrette, Romain Murenzi, and Bernard Piette. Image analysis with two-dimensional continuous wavelet transform. *Signal Processing*, 31(3):241 – 272, 1993.

[4] Tom M. Apostol. *Calculus: One-variable calculus, with an introduction to linear algebra*, volume 1, chapter Estimates for the error in Taylor's formula. Wiley, 2nd edition, 1967.

[5] Ron Appel, Thomas Fuchs, Piotr Dollár, and Pietro Perona. Quickly boosting decision trees - pruning underachieving features early. In *International Conference on Machine Learning*, volume 28, pages 594–602, June 2013.

[6] Roberto H. Bamberger and Mark J. T. Smith. A filter bank for the directional decomposition of images: theory and design. *IEEE Transactions on Signal Processing*, 40(4):882–893, April 1992.

[7] Rodrigo Benenson, Markus Mathias, Tinne Tuytelaars, and Luc Van Gool. Seeking the strongest rigid detector. In *IEEE Conference on Computer Vision and Pattern Recognition*, pages 3666–3673, June 2013.

[8] Rodrigo Benenson, Mohamed Omran, Jan Hosang, and Bernt Schiele. Ten years of pedestrian detection, what have we learned? In *European Conference on Computer Vision workshops*, September 2014.

[9] Garrick Brazil, Xi Yin, and Xiaoming Liu. Illuminating pedestrians via simultaneous detection & segmentation. In *IEEE International Conference on Computer Vision*, Venice, Italy, October 2017.

[10] Joan Bruna and Stéphane Mallat. Invariant scattering convolution networks. *IEEE Transactions on Pattern Analysis and Machine Intelligence*, 35(8):1872–1886, August 2013.

[11] Zhaowei Cai, Quanfu Fan, Rogério Schmidt Feris, and Nuno Vasconcelos. A unified multi-scale deep convolutional neural network for fast object detection. In *European Conference on Computer Vision*, pages 354–370, September 2016.

[12] Zhaowei Cai, Mohammad J. Saberian, and Nuno Vasconcelos. Learning complexity-aware cascades for deep pedestrian detection. In *IEEE International Conference on Computer Vision*, pages 3361–3369, December 2015.

[13] Emmanuel J. Candès and David L. Donoho. New tight frames of curvelets and optimal representations of objects with piecewise C^2 singularities. *Communications on Pure and Applied Mathematics*, 57(2):219–266, 2004.

[14] Emmanuel J. Candès and David L. Donoho. Continuous curvelet transform: I. Resolution of the wavefront set. *Applied and Computational Harmonic Analysis*, 19(2):162 – 197, 2005.

[15] Emmanuel J. Candès and David L. Donoho. Continuous curvelet transform: II. Discretization and frames. *Applied and Computational Harmonic Analysis*, 19(2):198 – 222, 2005.

[16] John Canny. A computational approach to edge detection. *IEEE Transactions on Pattern Analysis and Machine Intelligence*, 8(6):679–698, November 1986.

[17] Ole Christensen. *An Introduction to Frames and Riesz Bases*. Applied and numerical harmonic analysis. Birkhäuser Basel, 2003.

[18] Charles K. Chui. *An Introduction to Wavelets*. Wavelet analysis and its applications. Academic Press, 1992.

[19] Charles K. Chui. *An Introduction to Wavelets*, chapter Cardinal Spline-Wavelets. Wavelet analysis and its applications. Academic Press, 1992.

[20] Charles K. Chui and Jian-Zhong Wang. On compactly supported spline wavelets and a duality principle. *Transactions of the American Mathematical Society*, 330(2), 1992.

[21] Stephan Dahlke, Gitta Kutyniok, Peter Maass, Chen Sagiv, Hans-Georg Stark, and Gerd Teschke. The uncertainty principle associated with the continuous shearlet transform. *International Journal of Wavelets, Multiresolution and Information Processing*, 6(2):157–181, 2008.

[22] Navneet Dalal and Bill Triggs. Histograms of oriented gradients for human detection. In *IEEE Computer Society Conference on Computer Vision and Pattern Recognition*, volume 1, pages 886–893 vol. 1, June 2005.

[23] Ingrid Daubechies. *Ten Lectures on Wavelets*. Society for Industrial and Applied Mathematics, 1992.

[24] Jia Deng, Wei Dong, Richard Socher, Li-Jia Li, Kai Li, and Li Fei-Fei. Imagenet: A large-scale hierarchical image database. In *IEEE Conference on Computer Vision and Pattern Recognition*, pages 248–255, June 2009.

[25] M. N. Do and M. Vetterli. The contourlet transform: an efficient directional multiresolution image representation. *IEEE Transactions on Image Processing*, 14(12):2091–2106, December 2005.

[26] Piotr Dollár. Piotr's Computer Vision Matlab Toolbox (PMT). Online, http://vision.ucsd.edu/~pdollar/toolbox/doc/index.html, July 2017.

[27] Piotr Dollár, Ron Appel, Serge Belongie, and Pietro Perona. Fast feature pyramids for object detection. *IEEE Transactions on Pattern Analysis and Machine Intelligence*, 36(8):1532–1545, August 2014.

[28] Piotr Dollár, Serge Belongie, and Pietro Perona. The fastest pedestrian detector in the west. In *Proceedings of the British Machine Vision Conference*, pages 68.1–68.11, September 2010.

[29] Piotr Dollár, Zhuowen Tu, Pietro Perona, and Serge Belongie. Integral channel features. In *Proceedings of the British Machine Vision Conference*, pages 91.1–91.11, September 2009.

[30] Piotr Dollár, Christian Wojek, Bernt Schiele, and Pietro Perona. Pedestrian detection: A benchmark. In *IEEE Conference on Computer Vision and Pattern Recognition*, pages 304–311, June 2009.

[31] Piotr Dollár, Christian Wojek, Bernt Schiele, and Pietro Perona. Pedestrian detection: An evaluation of the state of the art. *IEEE Transactions on Pattern Analysis and Machine Intelligence*, 34(4):743–761, April 2012.

[32] Xianzhi Du, Mostafa El-Khamy, Jungwon Lee, and Larry S. Davis. Fused DNN: A deep neural network fusion approach to fast and robust pedestrian detection. In *IEEE Winter Conference on Applications of Computer Vision*, pages 953–961, March 2017.

[33] Richard James Duffin and Albert Charles Schaeffer. A class of nonharmonic fourier series. *Transactions of the American Mathematical Society*, 72(2):341–366, 1952.

[34] Miguel A. Duval-Poo, Francesca Odone, and Ernesto De Vito. Edges and corners with shearlets. *IEEE Transactions on Image Processing*, 24(11):3768–3780, 2015.

[35] Glenn Easley, Demetrio Labate, and Wang-Q Lim. Sparse directional image representations using the discrete shearlet transform. *Applied and Computational Harmonic Analysis*, 25:25–46, July 2008.

[36] Glenn R. Easley and Demetrio Labate. *Shearlets: Multiscale Analysis for Multivariate Data*, chapter Image Processing Using Shearlets, pages 283–325. Birkhäuser, Boston, MA, USA, 2012.

[37] Euro NCAP. 2020 roadmap. Online, www.euroncap.blob.core.windows.net/media/16472/euro-ncap-2020-roadmap-rev1-march-2015.pdf, March 2015.

[38] Jerome Friedman, Robert Tibshirani, and Trevor Hastie. Additive logistic regression: A statistical view of boosting (with discussion and a rejoinder by the authors). *Annals of Statistics*, 28(2):337–407, April 2000.

[39] Gartner Inc. Hype cycle for emerging technologies. Online, http://www.gartner.com, July 2017.

[40] David Gerónimo, Antonio M. López, Angel Domingo Sappa, and Thorsten Graf. Survey of pedestrian detection for advanced driver assistance systems. *IEEE Transactions on Pattern Analysis and Machine Intelligence*, 32(7):1239–1258, 2010.

[41] Ross Girshick. Fast R-CNN. In *IEEE International Conference on Computer Vision*, December 2015.

[42] Ross Girshick, Jeff Donahue, Trevor Darrell, and Jitendra Malik. Rich feature hierarchies for accurate object detection and semantic segmentation. In *IEEE Conference on Computer Vision and Pattern Recognition*, pages 580–587, June 2014.

[43] David E. Goldberg. *Genetic Algorithms*. Pearson Education, 2006.

[44] Pierre Goupillaud, Alexandre Grossmann, and Jean Morlet. Cycle-octave and related transforms in seismic signal analysis. *Geoexploration*, 23(1):85 – 102, 1984. Seismic Signal Analysis and Discrimination III.

[45] K. Gröchenig. *Foundations of Time-Frequency Analysis*. Applied and Numerical Harmonic Analysis. Birkhäuser, Boston, MA, USA, 2013.

[46] Philipp Grohs. Continuous shearlet frames and resolution of the wavefront set. *Monatshefte für Mathematik*, 164(4):393–426, December 2011.

[47] Philipp Grohs. Continuous shearlet tight frames. *Journal of Fourier Analysis and Applications*, 17(3):506–518, June 2011.

[48] Philipp Grohs, Sandra Keiper, Gitta Kutyniok, and Martin Schäfer. α-molecules. *Applied and Computational Harmonic Analysis*, 41(1):297 – 336, 2016.

[49] Alexandre Grossmann and Jean Morlet. Decomposition of hardy functions into square integrable wavelets of constant shape. *SIAM Journal on Mathematical Analysis*, 15(4):723–736, 1984.

[50] Kanghui Guo, Robert Houska, and Demetrio Labate. Microlocal analysis of singularities from directional multiscale representations. In Gregory E. Fasshauer and Larry L. Schumaker, editors, *Approximation Theory XIV: San Antonio 2013*, pages 173–196. Springer International Publishing, 2014.

[51] Kanghui Guo, Gitta Kutyniok, and Demetrio Labate. Sparse multidimensional representations using anisotropic dilation and shear operators. In *Wavelets and splines: Athens 2005*, Mod. Methods Math., pages 189–201. Nashboro Press, Brentwood, TN, USA, 2006.

[52] Kanghui Guo and Demetrio Labate. Optimally sparse multidimensional representation using shearlets. *SIAM Journal on Mathematical Analysis*, 39(1):298–318, 2007.

[53] Kanghui Guo and Demetrio Labate. Characterization and analysis of edges using the continuous shearlet transform. *SIAM on Imaging Sciences*, 2(3):959–986, 2009.

[54] Kanghui Guo, Demetrio Labate, and Wang-Q Lim. Edge analysis and identification using the continuous shearlet transform. *Applied and Computational Harmonic Analysis*, 27(1):24 – 46, 2009.

[55] Kanghui Guo, Demetrio Labate, Wang-Q Lim, Guido Weiss, and Edward Wilson. Wavelets with composite dilations. *Electronic Research Announcements of the American Mathematical Society*, 10(9):78–87, August 2004.

[56] Kanghui Guo, Demetrio Labate, Wang-Q Lim, Guido Weiss, and Edward Wilson. *Harmonic Analysis and Applications: In Honor of John J. Benedetto*, chapter The Theory of Wavelets with Composite Dilations, pages 231–250. Birkhäuser, Boston, MA, USA, 2006.

[57] Kanghui Guo, Demetrio Labate, Wang-Q Lim, Guido Weiss, and Edward Wilson. Wavelets with composite dilations and their MRA properties. *Applied and Computational Harmonic Analysis*, 20(2):202 – 236, 2006.

[58] Bin Han, Gitta Kutyniok, and Zuowei Shen. Adaptive multiresolution analysis structures and shearlet systems. *SIAM Journal on Numerical Analysis*, 49(5/6):1921–1946, 2006.

[59] David Harel and Michal Politi. *Modeling Reactive Systems with Statecharts: The Statemate Approach*. Computing McGraw-Hill. McGraw-Hill, 1998.

[60] Sören Häuser. *Shearlet Coorbit Spaces, Shearlet Transforms and Applications in Imaging*. PhD thesis, Technische Universität Kaiserslautern, 2014.

[61] Sören Häuser and Gabriele Steidl. Fast finite shearlet transform: a tutorial. *ArXiv*, abs/1202.1773, 2014.

[62] Kaiming He, Xiangyu Zhang, Shaoqing Ren, and Jian Sun. Delving deep into rectifiers: Surpassing human-level performance on imagenet classification. In *IEEE International Conference on Computer Vision*, pages 1026–1034, December 2015.

[63] Kaiming He, Xiangyu Zhang, Shaoqing Ren, and Jian Sun. Deep residual learning for image recognition. In *IEEE Conference on Computer Vision and Pattern Recognition*, June 2016.

[64] Gert Heinrich and Klaus Mairon. *Objektorientierte Systemanalyse*. Wirtschaftsinformatik kompakt. De Gruyter, 2008.

[65] Sergey Ioffe and Christian Szegedy. Batch normalization: Accelerating deep network training by reducing internal covariate shift. In *International Conference on Machine Learning*, pages 448–456, July 2015.

[66] Yangqing Jia, Evan Shelhamer, Jeff Donahue, Sergey Karayev, Jonathan Long, Ross Girshick, Sergio Guadarrama, and Trevor Darrell. Caffe: Convolutional architecture for fast feature embedding. In *ACM International Conference on Multimedia*, pages 675–678, November 2014.

[67] Nick Kingsbury. Image processing with complex wavelets. *Philosophical Transactions of the Royal Society of London A: Mathematical, Physical and Engineering Sciences*, 357(1760):2543–2560, 1999.

[68] Nick Kingsbury. Complex wavelets for shift invariant analysis and filtering of signals. *Applied and Computational Harmonic Analysis*, 10(3):234 – 253, 2001.

[69] Pisamai Kittipoom, Gitta Kutyniok, and Wang-Q Lim. Construction of compactly supported shearlet frames. *Constructive Approximation*, 35(1):21–72, October 2011.

[70] Morris Kline. *Calculus: An Intuitive and Physical Approach*, chapter Taylor's Formula. Dover Books on Mathematics. Dover Publications, 1998.

[71] Alex Krizhevsky. Learning multiple layers of features from tiny images. Technical report, Computer Science Department, University of Toronto, 2009.

[72] Alex Krizhevsky, Ilya Sutskever, and Geoffrey E. Hinton. Imagenet classification with deep convolutional neural networks. In *Advances in Neural Information Processing Systems*, pages 1097–1105, 2012.

[73] Gitta Kutyniok and Demetrio Labate. Construction of regular and irregular shearlet frames. *Journal of Wavelet Theory and Applications*, 1(1):1–10, 2007.

[74] Gitta Kutyniok and Demetrio Labate. *Shearlets: Multiscale Analysis for Multivariate Data*. Birkhäuser, Boston, MA, USA, 2012.

[75] Gitta Kutyniok and Demetrio Labate. *Shearlets: Multiscale Analysis for Multivariate Data*, chapter Introduction to Shearlets, pages 1–38. Birkhäuser, Boston, MA, USA, 2012.

[76] Gitta Kutyniok and Wang-Q Lim. Compactly supported shearlets are optimally sparse. *Journal of Approximation Theory*, 163(11):1564 – 1589, 2011.

[77] Gitta Kutyniok, Wang-Q Lim, and Rafael Reisenhofer. ShearLab 3D: Faithful digital shearlet transforms based on compactly supported shearlets. *ACM Transactions on Mathematical Software*, 42(1):5:1–5:42, January 2016.

[78] Gitta Kutyniok and Philipp Petersen. Classification of edges using compactly supported shearlets. *Applied and Computational Harmonic Analysis*, 42(2):245 – 293, 2017.

[79] Gitta Kutyniok, Morteza Shahram, and David L. Donoho. Development of a digital shearlet transform based on pseudo-polar FFT. In *Wavelets XIII*, volume 7446 of *SPIE Proceedings*, pages 74460B–74460B–13, 2009.

[80] Gitta Kutyniok, Morteza Shahram, and Xiaosheng Zhuang. Shearlab: A rational design of a digital parabolic scaling algorithm. *SIAM Journal on Imaging Sciences*, 5(4):1291–1332, 2012.

[81] Demetrio Labate, Wang-Q Lim, Gitta Kutyniok, and Guido Weiss. Sparse multidimensional representation using shearlets. In *Wavelets XI*, volume 5914 of *SPIE Proceedings*, pages 254–262, August 2005.

[82] Yann Lecun, Léon Bottou, Yoshua Bengio, and Patrick Haffner. Gradient-based learning applied to document recognition. *Proceedings of the IEEE*, 86(11):2278–2324, November 1998.

[83] John M. Lee. *Introduction to Smooth Manifolds*, chapter Smooth Maps, pages 30–59. Springer-Verlag New York, 1st edition, 2003.

[84] Jianan Li, Xiaodan Liang, ShengMei Shen, Tingfa Xu, and Shuicheng Yan. Scale-aware Fast R-CNN for pedestrian detection. *IEEE Transactions on Multimedia*, 20(4):985–996, April 2018.

[85] Song Li and Yi Shen. Shearlet frames with short support. *ArXiv*, abs/1101.4725, 2011.

[86] Wang-Q Lim. The discrete shearlet transform: A new directional transform and compactly supported shearlet frames. *IEEE Transactions on Image Processing*, 19(5):1166–1180, May 2010.

[87] Wang-Q Lim. Nonseparable shearlet transform. *IEEE Transactions on Image Processing*, 22(5):2056–2065, May 2013.

[88] Erik Lindholm, John Nickolls, Stuart Oberman, and John Montrym. NVIDIA Tesla: A Unified Graphics and Computing Architecture. *IEEE Micro*, 28(2):39–55, March 2008.

[89] Wei Liu, Dragomir Anguelov, Dumitru Erhan, Christian Szegedy, Scott Reed, Cheng-Yang Fu, and Alexander C. Berg. SSD: Single Shot multibox Detector. In *European Conference on Computer Vision*, pages 21–37. Springer International Publishing, October 2016.

[90] Stéphane Mallat. *A Wavelet Tour of Signal Processing*. San Diego : Academic Press, 1999.

[91] Stephane Mallat and Sifen Zhong. Characterization of signals from multiscale edges. *IEEE Transactions on Pattern Analysis and Machine Intelligence*, 14(7):710–732, July 1992.

[92] Stéphane Mallat. Group invariant scattering. *Communications on Pure and Applied Mathematics*, 65(10):1331–1398, October 2012.

[93] Yves Meyer. *Oscillating Patterns in Image Processing and Nonlinear Evolution Equations*. American Mathematical Society, 2001.

[94] Microsoft Corporation. Microsoft Robotics Developer Studio (MRDS). Online, http://www.microsoft.com/robotics, July 2017.

[95] Gradimir V. Milovanović and Zlatko Udovičić. Calculation of coefficients of a cardinal b-spline. *Applied Mathematics Letters*, 23(11):1346 – 1350, 2010.

[96] Woonhyun Nam, Piotr Dollár, and Joon Hee Han. Local decorrelation for improved pedestrian detection. In *Advances in Neural Information Processing Systems 27: Annual Conference on Neural Information Processing Systems*, pages 424–432, December 2014.

[97] NVIDIA Corporation. Jetson TK1 development kit user guide. Online, http://developer.download.nvidia.com/embedded/jetson/TK1/docs/2_GetStart/Jeston_TK1_User_Guide.pdf, November 2014.

[98] NVIDIA Corporation. NVIDIA Jetson TK1 developer kit. Online, http://www.nvidia.com/object/jetson-tk1-embedded-dev-kit.html, September 2016.

[99] Eshed Ohn-Bar and Mohan M. Trivedi. To boost or not to boost? On the limits of boosted trees for object detection. In *IEEE International Conference on Pattern Recognition*, December 2016.

[100] Open Source Robotics Foundation. Robot Operating System (ROS). Online, http://www.ros.org, September 2016.

[101] Wanli Ouyang, Hui Zhou, Hongsheng Li, Quanquan Li, Junjie Yan, and Xiaogang Wang. Jointly learning deep features, deformable parts, occlusion and classification for pedestrian detection. *IEEE Transactions on Pattern Analysis and Machine Intelligence*, 40:1874–1887, August 2018.

[102] Edouard Oyallon, Eugene Belilovsky, and Sergey Zagoruyko. Scaling the scattering transform: Deep hybrid networks. In *IEEE International Conference on Computer Vision*, October 2017.

[103] Edouard Oyallon and Stéphane Mallat. Deep roto-translation scattering for object classification. In *IEEE Conference on Computer Vision and Pattern Recognition*, pages 2865–2873, June 2015.

[104] Edouard Oyallon, Stéphane Mallat, and Laurent Sifre. Generic deep networks with wavelet scattering. *ArXiv*, abs/1312.5940, 2013.

[105] Sakrapee Paisitkriangkrai, Chunhua Shen, and Anton van den Hengel. Strengthening the effectiveness of pedestrian detection with spatially pooled features. In *European Conference on Computer Vision*, September 2014.

[106] Constantine Papageorgiou and Tomaso Poggio. A trainable system for object detection. *International Journal of Computer Vision*, 38(1):15–33, June 2000.

[107] Lienhard Pfeifer. Shearlet features for pedestrian detection. *Journal of Mathematical Imaging and Vision*, July 2018. https://doi.org/10.1007/s10851-018-0834-9.

[108] Michael J. Quinn. *Parallel Programming in C with MPI and OpenMP*. McGraw-Hill Higher Education. McGraw-Hill Higher Education, 2004.

[109] Shaoqing Ren, Kaiming He, Ross Girshick, and Jian Sun. Faster R-CNN: Towards real-time object detection with region proposal networks. *IEEE Transactions on Pattern Analysis and Machine Intelligence*, 39(6):1137–1149, June 2017.

[110] Dominik Scherer, Andreas Müller, and Sven Behnke. Evaluation of pooling operations in convolutional architectures for object recognition. In *International Conference on Artificial Neural Networks*, pages 92–101, September 2010.

[111] William Robson Schwartz, Ricardo Dutra da Silva, Larry S. Davis, and Hélio Pedrini. A novel feature descriptor based on the shearlet transform. In *IEEE International Conference on Image Processing*, pages 1033–1036, September 2011.

[112] E. P. Simoncelli, W. T. Freeman, E. H. Adelson, and D. J. Heeger. Shiftable multiscale transforms. *IEEE Transactions on Information Theory*, 38(2):587–607, March 1992.

[113] Karen Simonyan and Andrew Zisserman. Very Deep Convolutional Networks for Large-Scale Image Recognition. *ArXiv*, abs/1409.1556, September 2014.

[114] Statistisches Bundesamt. Unfallentwicklung auf deutschen Straßen 2015. Online, https://www.destatis.de/DE/Publikationen/Thematisch/TransportVerkehr/Verkehrsunfaelle/, July 2016.

[115] Christian Szegedy, Wei Liu, Yangqing Jia, Pierre Sermanet, Scott Reed, Dragomir Anguelov, Dumitru Erhan, Vincent Vanhoucke, and Andrew Rabinovich. Going deeper with convolutions. In *IEEE Conference on Computer Vision and Pattern Recognition*, June 2015.

[116] Christian Szegedy, Wojciech Zaremba, Ilya Sutskever, Joan Bruna, Dumitru Erhan, Ian Goodfellow, and Rob Fergus. Intriguing properties of neural networks. In *International Conference on Learning Representations*, April 2014.

[117] Yonglong Tian, Ping Luo, Xiaogang Wang, and Xiaoou Tang. Deep learning strong parts for pedestrian detection. In *IEEE International Conference on Computer Vision*, pages 1904–1912, December 2015.

[118] Paul A. Viola and Michael J. Jones. Rapid object detection using a boosted cascade of simple features. In *IEEE Computer Society Conference on Computer Vision and Pattern Recognition*, pages 511–518, December 2001.

[119] Anders Vretblad. *Fourier Analysis and Its Applications*, volume 223 of *Graduate Texts in Mathematics*. Springer, 2003.

[120] E.W. Weisstein. *CRC Concise Encyclopedia of Mathematics*. CRC Press, 2nd edition, 2002.

[121] Paul John Werbos. *Beyond Regression: New Tools for Prediction and Analysis in the Behavioral Sciences*. PhD thesis, Harvard University, Cambridge, MA, USA, 1974.

[122] Thomas Wiatowski and Helmut Bölcskei. Deep convolutional neural networks based on semi-discrete frames. In *IEEE International Symposium on Information Theory*, pages 1212–1216, June 2015.

[123] Thomas Wiatowski and Helmut Bölcskei. A mathematical theory of deep convolutional neural networks for feature extraction. *IEEE Transactions on Information Theory*, 64(3):1845–1866, March 2018.

[124] Deyun Yang and Xingwei Zhou. Irregular wavelet frames on $L^2(\mathbb{R}^n)$. *Science in China Series A-Mathematics*, 48(277).

[125] Sheng Yi, Demetrio Labate, Glenn R. Easley, and Hamid Krim. A shearlet approach to edge analysis and detection. *IEEE Transactions on Image Processing*, 18(5):929–941, 2009.

[126] Fisher Yu and Vladlen Koltun. Multi-scale context aggregation by dilated convolutions. In *International Conference on Learning Representations*, May 2016.

[127] Liliang Zhang, Liang Lin, Xiaodan Liang, and Kaiming He. Is Faster R-CNN doing well for pedestrian detection? In *European Conference on Computer Vision*, pages 443–457, September 2016.

[128] Shanshan Zhang, Christian Bauckhage, and Armin B. Cremers. Informed haar-like features improve pedestrian detection. In *IEEE Conference on Computer Vision and Pattern Recognition*, pages 947–954, June 2014.

[129] Shanshan Zhang, Rodrigo Benenson, Mohamed Omran, Jan Hosang, and Bernt Schiele. How far are we from solving pedestrian detection? In *IEEE Conference on Computer Vision and Pattern Recognition*, pages 1259–1267, June 2016.

[130] Shanshan Zhang, Rodrigo Benenson, and Bernt Schiele. Filtered channel features for pedestrian detection. In *IEEE Conference on Computer Vision and Pattern Recognition*, pages 1751–1760, June 2015.

Index